9/2007

Texts and Monographs in Computer Science

Editor
David Gries

Advisory Board
F. L. Bauer
J. J. Horning
R. Reddy
D. C. Tsichritzis
W. M. Waite

Texts and Monographs in Computer Science

Adaptive Signal Processing

Theory and Applications

S. Thomas Alexander

With 42 Illustrations

Springer-Verlag New York Berlin Heidelberg
London Paris Tokyo

S. Thomas Alexander
Department of Electrical and
    Computer Engineering
North Carolina State University
Raleigh, NC 27695-7911
U.S.A.

Series Editor

David Gries
Department of Computer Science
Cornell University
Upson Hall
Ithaca, NY 14853
U.S.A.

Library of Congress Cataloging in Publication Data
Alexander, S. Thomas.
    Adaptive signal processing.
    (Texts and monographs in computer science)
    Bibliography: p.
    Includes index.
    1. Adaptive signal processing.   I. Title.
II. Series.
TK5102.5.A424   1986   621.38′043   86-13956

Typeset by Asco Trade Typesetting Ltd., Hong Kong.
Printed and bound by R.R. Donnelley & Sons, Harrisonburg, Virginia.
Printed in the United States of America.

9 8 7 6 5 4 3 2 1

ISBN 0-387-96380-4 Springer-Verlag New York Berlin Heidelberg
ISBN 3-540-96380-4 Springer-Verlag Berlin Heidelberg New York

Preface

The creation of the text really began in 1976 with the author being involved with a group of researchers at Stanford University and the Naval Ocean Systems Center, San Diego. At that time, adaptive techniques were more laboratory (and mental) curiosities than the accepted and pervasive categories of signal processing that they have become. Over the last 10 years, adaptive filters have become standard components in telephony, data communications, and signal detection and tracking systems. Their use and consumer acceptance will undoubtedly only increase in the future.

The mathematical principles underlying adaptive signal processing were initially fascinating and were my first experience in seeing applied mathematics work for a paycheck. Since that time, the application of even more advanced mathematical techniques have kept the area of adaptive signal processing as exciting as those initial days. The text seeks to be a bridge between the open literature in the professional journals, which is usually quite concentrated, concise, and advanced, and the graduate classroom and research environment where underlying principles are often more important.

In that spirit, this text will be most beneficially used as an introductory tool for anyone interested in learning the fascinating field of adaptive signal processing. Most of the intended audience will be seniors and graduate students in electrical engineering or computer science, although the practicing engineer "gearing up" to work on product development using adaptive techniques will also find the text useful. An understanding of linear systems, digital signal processing, and matrix algebra approximately equivalent to that of an undergraduate electrical engineering curriculum is adequate. The text has been used for a graduate course in adaptive signal processing at North Carolina State University, in which students from a wide variety of backgrounds have actively participated.

The main distinction between this text and others that have appeared on the subject is the inclusion in this text of the timely subject of vector space approaches to fast adaptive filtering. This is currently one of the most active areas of research in signal processing, but the mathematical sophistication required to understand the open literature in the area has been formidable. This text develops the vector space approach through the liberal use of geometrical analogies, which encompasses Chapters 9–11. In so doing, the vector space approach becomes actually very easy to understand and possesses a great deal of simple elegance. After completing these chapters, the reader will be well prepared to tackle some of the more specific research problems associated with fast adaptive techniques. The book can be an effective text for a one-semester course in adaptive signal processing or as a reference for the researcher (academic or industrial) to absorb material at his own pace. Problems that have survived the classroom experience are included at the end of the chapters.

This text is approximately the same process by which I became familiar with the different areas of adaptive signal processing. Much of the original material was literally "back of the envelope" information from hearing conference talks and informal discussions. Other portions had their beginning as notes scribbled in the margin of books or papers when something finally jelled.

As with any book, the contributions of many people over many years were instrumental to the entire process. Specifically, I would like to thank the following: Bob Plemmons of North Carolina State for sharing his mastery of linear algebra; Lloyd Griffiths of USC for long runs during which philosophies were discussed; Ed Satorius of JPL and Joel Trussell of North Carolina State for consistently providing honest and, therefore, valuable technical evaluation and discussion on both this text and the Big Picture; John Cioffi of Stanford for his contributions to my own understanding of geometrical approaches and fast adaptive techniques; Nino Masnari, Chairman of the Electrical and Computer Engineering Department at North Carolina State, for helping to foster the professional environment that allowed the time and resources to develop this text; and the editorial staff at Springer-Verlag for providing the assistance and encouragement I needed to successfully complete the manuscript. Additionally, for their unsung efforts in reading and debugging the original drafts and homework problems, I would like to thank the following at North Carolina State: Gary Ybarra, Glenda Poston, Zong Rhee, Daehoon Kim, and Randy Avent. Special thanks are also extended to Peggy Ball, Liz Story, George Winston, and Red Ryder.

Finally, the city of Boston and the season of Winter had a lot to do with the whole process.

Raleigh, NC   S.T. ALEXANDER

Contents

CHAPTER 1
Introduction

1.1 Signal Processing in Unknown Environments

Many everyday problems encountered in communications and signal processing involve removing noise and distortion due to physical processes that are time varying or unknown or possibly both. These types of processes represent some of the most difficult problems in transmitting and receiving information. The area of adaptive signal processing techniques provides one approach for removing distortion in communications, as well as extracting information about unknown physical processes. A short consideration of some of these problems shows that distortion is often present regardless of whether the communication is conversation between people or data between physical devices.

For example, a common problem in long distance telephone communications is the creation of echoes due to impedance mismatches on the network. This has an extremely annoying effect on the persons using the telephone link, and can degrade the quality of communication such that the conversation is rendered unsatisfactory.

Another example is that of computers exchanging data over physical communications channels. Many channels are well conditioned and deliver the original transmitted pulses undistorted to the receiver. However, many channels are poorly conditioned and distort the received digital pulses to such a degree that data decision devices would make far too many errors to provide a useful service.

While there are numerous other examples, these two are sufficient to illustrate some of the main reasons for needing adaptive signal processors. In the first example above, the impedance mismatch is usually unknown. That is, the sheer number of local telephone lines that must be accessed effectively prohibits the impedance for any one local line to be accurately matched to the long distance link. Even if the resources were available to match the impedance of each local line to the long distance link, there is still the problem that due to aging, inaccurate component values, moisture, etc., the impedance of each local line may be time varying. Therefore, attempts to build a single processor that has the flexibility of addressing all these time-varying and unknown phenomena require adaptivity in the processor. Such adaptive signal processing devices are widely in use now and are known as adaptive echo cancellers.

In the second example above concerning data transmission, the unknown and/or time-varying segments is the communications channel itself. For example, with the proliferation of mobile radios, there has emerged the possibility of transmitting and receiving data from highly mobile stations to a central computer database. Consider the case in which the transmitter is mobile (i.e., located in a car). Since the data is encoded and sent over the atmospheric radio channel, the propagation path between the transmitter and fixed receiver is changing, sometimes quite rapidly. Therefore, such considerations as data symbol timing, power of received signal, and propagation loss, for example, are time varying and unknown to the system. Once again, designing a fixed parameter system to handle the wide range of *possible* values of these parameters could render the system performance unacceptable for certain specifically encountered situations.

The common element in each of the preceding problems, and indeed in most of the applications of adaptive signal processing, is that some element of the problem is unknown and must therefore be learned, or some component of the system is changing in an unknown manner and therefore must be tracked. Quite frequently, both of these problems are resident in the applications of adaptive signal processing.

1.2 Two Examples

Two general examples were discussed in the previous section. This section will discuss two additional examples in more detail and provide a mathematical framework for adaptive signal processing. These two applications examples will be used throughout the text to illustrate new concepts and to investigate the performance characteristics of various adaptive methods. The first example is known as systems identification and the second will be referred to as linear prediction.

Systems identification

The first example concerns systems identification, which is used quite frequently in controls and communications work. Consider the case of Figure 1.1, in which it is desired to learn the structure of an unknown system from a

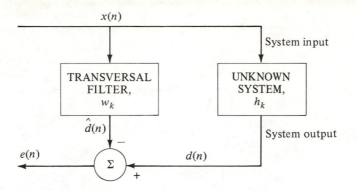

Figure 1.1   Identification of unknown system, h_k, using transversal filter, w_k.

knowledge of its input $x(n)$ and output $d(n)$. For example, it may be necessary to determine if any of the system parameters are approaching critical values. Although there are several way of quantifying the "knowledge of a system," the system impulse response, h_k, is often used. From elementary linear systems,

$$d(n) = \sum_{k=0}^{n} h_k x(n-k), \tag{1.2.1}$$

where both the input signal and the impulse response have been assumed to be casual. In this problem, the true impulse response h_k is unknown and must be obtained.

The form of (1.2.1) suggests the following approach to "learning" the h_k. Assume the h_k form a finite impulse response of no more than N samples in duration, counting h_0. Then,

$$d(n) = \sum_{k=0}^{N-1} h_k x(n-k). \tag{1.2.2}$$

This is the true system output $d(n)$. Using the form suggested by (1.2.2), a prediction of $d(n)$, denoted as $\hat{d}(n)$, may be made using a set of filter coefficients w_k:

$$\hat{d}(n) = \sum_{k=0}^{N-1} w_k x(n-k). \tag{1.2.3}$$

Strictly speaking, (1.2.3) is an estimation of the signal $d(n)$. However, much of the current literature refers to a form such as (1.2.3) as a prediction of $d(n)$, and this terminology will be used in this book.

If each chosen w_k is "close" to each true h_k, then the prediction error,

$$e(n) = d(n) - \hat{d}(n) \tag{1.2.4}$$

should be small in magnitude. In systems identification, the rationale is that if $\hat{d}(n) \approx d(n)$, then $w_k \approx h_k$. Therefore, minimizing some measure of $e(n)$ should force the individual w_k to approach the individual h_k, thus identifying the

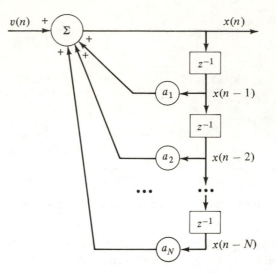

Figure 1.2   Information signal $x(n)$ produced by Nth order autoregressive (AR) process.

system. In Chapter 2, the mean square error will be seen to be a natural measure that is amenable to mathematical analysis, as well.

Linear prediction

The second example is very common in speech analysis and telecommunications. Consider Figure 1.2, which shows a model of a process that generates an information signal. The output of the model is the current sample $x(n)$, and it is easy to see that $x(n)$ is given by

$$x(n) = \sum_{k=1}^{N} a_k x(n - k) + v(n), \qquad (1.2.5)$$

where the a_k are the unknown system parameters and $v(n)$ is an unknown random excitation sequence. In statistical literature, (1.2.5) is called an Nth order autoregressive (AR) process. The model structure in (1.2.5) is analogous to a signal produced through a series of reflecting/transmitting media. This is a very good model for speech produced in the vocal tract, seismic signals propagating through a layered earth, and certain types of electromagnetic reflections in radar applications.

The form of the model in (1.2.5) is an approach similar to the preceding example for "learning" the generating parameters of the unknown system. That is, form a linear prediction of $x(n)$ based upon the N most recent $x(n - 1)$, ..., $x(n - N)$:

$$\hat{x}(n) = \sum_{k=1}^{N} w_k x(n - k). \qquad (1.2.6)$$

Since the excitation sequence $v(n)$ is unknown, it clearly cannot be used in the prediction (1.2.6). The prediction error then becomes

$$e(n) = x(n) - \hat{x}(n). \tag{1.2.7}$$

In this text, predicting a signal sample based upon previous values of the same signal will be called the purely linear prediction problem or, more simply, the linear prediction problem. Since the excitation sequence $v(n)$ is unknown, it clearly cannot be used in the prediction (1.2.6).

Once more, the rationale is that if the mean square of $e(n)$ is small, then the w_k of the filter will be forced to the a_k of the signal model. Sometimes in speech communications problems, the object is to identify and transmit the a_k. Other telecommunications applications might transmit information about $e(n)$. These applications will be discussed in more detail as they naturally arise during the course of the text.

In either of the preceding examples, the filter coefficients w_k may be constant if the unknown process is at least statistically stationary. This will be referred to as the fixed filter case. However, if the system to be identified in (1.2.1) or the information process in (1.2.5) has parameters that change in an unknown manner, then adaptivity of the filter coefficients is necessary. These two examples of systems identification and linear prediction will be referred to frequently throughout the text in more detail. For additional applications of adaptive filtering, as well as other approaches to the subject of adaptive signal processing, the reader is directed to references [1] through [9].

1.3 Outline of the Text

One purpose of this text is to develop in a cohesive, structured approach some of the more useful and promising adaptive signal processing techniques. This approach will be designed to accurately display the common foundations of these methods, but it will also illuminate the differences between the candidate adaptive approaches. This should allow the systems designer the tools for selecting the appropriate approach to the problem at hand, given the engineering constraints of memory, speed, and cost.

Another purpose is to provide the reader with a well-founded physical and geometrical understanding of the adaptive signal processing methods. It is sometimes tempting to launch immediately into mathematical derivations without providing the often necessary physical understanding of the adaptive processes. However, this text strives to demonstrate physical or geometrical interpretation whenever possible, thus providing another tool for understanding adaptive methods. To this end, a general outline of the text is as follows.

Since the concept of minimum mean square error (MSE) is prerequisite to understanding much of modern adaptive signal processing, Chapter 2 develops a foundation for determining minimum MSE filters. Necessary probabilistic and statistical concepts are also introduced as they are needed in this chapter. In addition, the very important normal equations are derived in Chapter 2. The concept of a "bowl-shaped" error surface is developed, as well as the geometrical analogy of locating the minimum of this surface.

Having thus formulated the minimum MSE problem, Chapter 3 then explores one very important method of solving the normal equations that result in the very important application of linear prediction filtering. This leads to the Durbin recursion, which in turn leads to the lattice filter structure. This lattice structure will be seen to be very useful in many approaches to linear prediction and adaptive signal processing and provides an alternative to the transversal filter implementation.

Another important iterative approach for solving the normal equations is developed in Chapter 4. This is the gradient-based technique known as the method of steepest descent. A derivation of the convergence properties of this method is presented, as well as the valuable geometrical analogy of "finding the bottom of the bowl."

Chapter 5 then makes the transition to the least mean squares (LMS) algorithm, which computes an approximation to the method of steepest descent. The analytical convergence properties of the LMS algorithm are developed in detail in this chapter. Additionally, the differences between LMS and steepest descent are discussed in detail, as well as considerations for using the LMS algorithm in actual systems.

Two application examples, known as systems identification and linear prediction, are developed in Chapters 2–5. Chapter 6 then provides some additional applications of adaptive filtering, which give more insight into actual systems usage. The applications of Chapter 6 are all done using the popular LMS algorithm.

Chapter 7 then discusses some adaptive approaches based upon Durbin's algorithm that converge more rapidly than LMS for most applications. Known collectively as gradient-based lattice techniques, these approaches create sets of orthogonal signals from the acquired data signal, which are then used in efficient updating algorithms.

The modern area of recursive least squares (RLS) adaptive filters is next introduced in Chapter 8. The specific method of Chapter 8 is the regular RLS method, which lays the foundation for understanding the extremely rapidly converging techniques of modern least squares filtering. The RLS method investigated is different from the gradient-based methods examined thus far, in that it computes the optimal least squares prediction at every point in time. Gradient-based methods, such as LMS, are only optimal at convergence. This added performance capability is not without cost, however, since the regular RLS requires substantially more computation than LMS.

However, there are newly derived methods of reducing the required computations in RLS filters, which are collectively known as fast RLS techniques. Chapter 9 recasts the RLS problem by structuring it as a minimization problem in a Hilbert space, which has some very beneficial geometrical interpretations. This leads to the very powerful vector space approach to fast adaptive filters, which can be applied to either the fast lattice or fast transversal implementations.

Chapter 10 then applies these vector space concepts to the derivation of the fast least squares lattice (LSL) filter for linear prediction. The geometrical interpretations of the LSL are emphasized in this chapter.

Chapter 11 then applies the geometrical concepts of Chapter 9 and derives the least squares fast transversal filter (FTF), which is the transversal counterpart of the LSL implementation in Chapter 10. The FTF has the fewest arithmetic operations per time update of any least squares method derived to date. In Chapters 10 and 11, it will be seen that the LSL and FTF approaches usually converge much more rapidly than the gradient-based adaptive filters. Since their computational complexity is comparable to LMS, they are indeed worthy of consideration for many modern applications.

REFERENCES

1. S. Haykin, *Introduction to Adaptive Filters*, Macmillan, New York, 1984.
2. R.A. Monzingo and T.W. Miller, *Introduction to Adaptive Arrays*, Wiley-Interscience, New York, 1980.
3. F. Hsu and A.A. Giordano, *Least Squares Signal Processing*, John Wiley & Sons, New York, 1984.
4. B. Widrow and S.D. Stearns, *Adaptive Signal Processing*, Prentice-Hall, Englewood Cliffs, NJ, 1985.
5. J.G. Proakis, *Digital Communications*, McGraw-Hill, New York, 1984.
6. S.J. Orfanidis, *Optimum Signal Processing*, Macmillan, New York, 1985.
7. G.C. Goodwin and K.S. Sin, *Adaptive Filtering, Prediction, and Control*, Prentice-Hall, Englewood Cliffs, NJ, 1984.
8. M.L. Honig and D.G. Messerschmitt, *Adaptive Filters*, Kluwer Academic Publishers, Hingham, MA, 1984.
9. B.D.O. Anderson and J.B. Moore, *Optimal Filtering*, Prentice-Hall, Englewood Cliffs, NJ, 1979.

CHAPTER 2
The Mean Square Error (MSE) Performance Criteria

2.1 Introduction

Adaptive signal processing algorithms generally attempt to optimize a performance measure that is a function of the unknown parameters to be identified. The most pervasive of these performance measures are based upon squared prediction errors, although the specific prediction error used in adaptation often depends upon the particular algorithm. Two broad categories of adaptive signal processing methods are: (1) stochastic and (2) exact. The latter category refers to adaptive filters based upon the actual or exact data signals acquired. The recursive least squares techniques comprising Chapters 8–11 are examples of these exact techniques, and investigation of those techniques will be deferred until the later chapters.

The former category of adaptive techniques known collectively as stochastic methods are based upon derivations using the statistical properties of the data signals. The primary statistical measure used is the ensemble average, or mean, of a squared prediction error function, and this has evolved into widespread use of the mean squared prediction error as a performance measure. Often this is shortened to simply the mean square error (MSE).

Many of the properties of minimum MSE filters and the MSE surface are derived using basic linear algebra techniques, such as eigenvalue and eigenvector analysis. Excellent texts on linear algebra at the introductory level are those by Anton [1] and Strang [2]. Another text that is particularly strong in geometrical interpretations is by Moore [3]. At the somewhat more advanced level is the text by Noble and Daniel [4], which is written largely from an engineering and physical science standpoint. As a result, the development of eigenvalues and eigenvectors is superb and gives excellent physical analogies. A more expanded treatment of physical considerations is given in an earlier edition by Noble [5]. The book by Bellman [6] is a somewhat more mathematically oriented text, but is still very good for the engineering student and contains a number of worked examples. For the active researcher in mean square prediction, filtering, and estimation, the text by Fadeev and Fadeeva [7] offers insight into both the theoretical and computational aspects of matrix and eigensystem analysis. Additionally, Golub and Van Loan [17] is an excellent text for advanced matrix analysis and computational considerations.

Linear prediction techniques using the MSE criteria have been applied extensively in the speech research community. Many of the speech applications, as well as early general theoretical work, are contained in the book by Markel and Gray [8]. An excellent journal article that displays the flexibility of linear prediction to diverse applications is that by Makhoul [9]. Since the autocorrelation function plays such an important role in linear prediction and MSE analysis, the paper by Markel and Gray [10] analyzes in detail its impact. The text by Rabiner and Schafer [11] develops approximations to the autocorrelation function and applies them to the problem of linear prediction of speech. Finally, the excellent text by Jayant and Noll [12] illustrates the use of random process theory in many techniques chosen from the areas of speech and image coding.

Minimization of the MSE is the objective of many currently used adaptive methods, such as the least mean square (LMS) algorithm and the gradient lattice method. These techniques are the topics of Chapters 4–7. In the current chapter, the basics of minimizing the MSE using the techniques of linear prediction filtering are introduced. Section 2.2 develops the concept of a quadratic error surface, which has the simple geometrical property of a single, or global, minimum. Section 2.3 then discusses some additional properties of the error surface, which will be useful and important for relating the error surface to physical phenomena. Section 2.4 then derives the relation known as the normal equations, which defines the location of this global minimum. Section 2.5 then concludes with a discussion of some of the geometrical properties of the error surface, which aids in understanding the dynamic properties of the adaptive methods.

2.2 Mean Square Error (MSE) and MSE Surface

A rationale for the MSE as a performance measure is perhaps best illustrated by example. In the general case, it is sometimes desired to predict the current sample of one signal, $d(n)$, using samples of a second signal, $x(n)$. An example of this case was the systems identification application of Figure 1.1. In this text, $d(n)$ will be called the desired signal and $x(n)$ will be called the data signal, since it is desired to predict $d(n)$ using $x(n)$. Ideally, the prediction filter output should be exactly $d(n)$, but in actual systems, usually only a close approximation is possible. If the prediction filter output is not exactly equal to the desired signal, then their is some prediction error generated at time n. This prediction error is defined as the difference, $e(n)$, between the prediction filter output $\hat{d}(n)$ and the desired signal. Therefore, in this situation,

$$e(n) = d(n) - \hat{d}(n). \tag{2.2.1}$$

In most systems, a "large" value of negative error is as detrimental as a large value of positive error. For that reason, some performance measure that equally weighs large positive and negative errors is desirable. One immediate candidate is the absolute value, $|e(n)|$. However, this measure often leads to intractable mathematical results when attempting to analyze the resulting algorithms. The measure of squared prediction error, $e^2(n)$, lends itself much better to analytical work and has therefore become very widely used. The squared error follows directly from (2.2.1):

$$e^2(n) = [d(n) - \hat{d}(n)]^2. \tag{2.2.2}$$

Very often the desired signal may be the result of a random or *stochastic* process, due to random noise being added to an information signal or due to the information signal itself being generated by a random process. For example, in the application of linear prediction of speech from Section 1.2, the speech signal itself was seen to be a stochastic signal. If the desired signal is stochastic, then the error signal is also stochastic. The statistical property of the squared error signal, which is well suited as a performance measure, is the ensemble average or mean. The MSE will be defined as the ensemble average or expectation of the squared error sequence. Since this is an ensemble average and not a time average, it is admissible to define the MSE as a function of time, denoted as

$$\varepsilon(n) = E\{e^2(n)\}, \tag{2.2.3}$$

where $E\{\cdot\}$ is the expectation operator [11, 13, 14]. One area of active research is the examination of the convergence characteristics of adaptive filters via analytical methods, and an indicator of convergence behavior is $\varepsilon(n)$ as the adaptive filter "adapts." The variable n in $\varepsilon(n)$ is then analogous to the number of adaptive filter iterations that have been performed. In other cases, the problem of interest might be the steady-state characteristics after the filter has converged. This steady-state MSE will sometimes be simply denoted as ε without a time argument.

Linear minimum MSE filtering

This text examines adaptive filters that compute their output in a linear fashion. That is, given a set of N filter coefficients, $w_i(n)$, and a data sequence, $x(n)$, then the prediction of the desired signal may be computed in the linear form

$$\hat{d}(n) = \sum_{i=0}^{N-1} w_i(n)x(n-i) = \mathbf{w}_N^T(n)\mathbf{x}_N(n), \qquad (2.2.4a)$$

where

$$\mathbf{w}_N^T(n) = [w_0(n), \ldots, w_{N-1}(n)], \qquad (2.2.4b)$$

$$\mathbf{x}_N^T(n) = [x(n), \ldots, x(n-N+1)] \qquad (2.2.4c)$$

are the N components of weight coefficients and data samples, respectively. The subscript N in (2.2.4) denotes the number of components in the vector and n denotes the time argument of its first component. Employing a linear filter to form this prediction leads to many benefits, not the least of which are ease of implementation and mathematical analysis. The use of linear minimum MSE filtering is a rich and broadly applicable discipline, but by necessity the current investigation will be limited to those portions necessary for understanding adaptive signal processing. The interested reader is referred to the excellent tutorial by Makhoul [9], we well as the texts by Markel and Gray [8] and Rabiner and Schafer [11], for more detailed developments. Additionally, the text by Schwartz and Shaw [16] develops a number of practical methods for stochastic signal processing and provides many useful physical applications.

The stochastic adaptive algorithms to be examined in this text will all stem logically from the choice of MSE as a performance measure, therefore, it is natural to begin by examining the consequences and impact of using the MSE as a performance measure. While the general goal of this text is to study and understand the uses of adaptive filtering, there is a wealth of qualitative insight to be gained about the behavior of adaptive filters by first examining properties of linear minimum MSE filters and the MSE criterion. This is the area of classical minimum MSE prediction and estimation, and an understanding of this area will greatly aid the mathematical and conceptual transition to adaptive filtering. The notations, definitions, and many of the analytical techniques to be employed for adaptive techniques have their basis in the classical MSE fixed filter results. The first concept to be considered is the MSE surface.

For each value of filter coefficient vector $\mathbf{w}_N(n)$ selected in (2.2.4), there results a corresponding value of mean square prediction error. Since this MSE is therefore a function of $\mathbf{w}_N(n)$, this functional dependency may be denoted as $\varepsilon(\mathbf{w}_N)$. In this section, the functional relation between the weight vector $\mathbf{w}_N(n)$ and the MSE $\varepsilon(\mathbf{w}_N)$ is derived. Furthermore, this initial investigation will be limited to the case of weight vectors which are constant with respect to time and many thus be denoted as $\mathbf{w}_N$. Additionally, the signals $d(n)$ and $x(n)$ will be assumed to be statistically stationary [13]. After deriving the mathematical results, a systems identification example will be studied to provide physical analogies for the analytical results.

With the definition of the prediction error $e(n)$ from (2.2.1) and $\hat{d}(n)$ from (2.2.4), the expression for the mean square prediction error becomes

$$\varepsilon(\mathbf{w}_N) = E\{e^2(n)\} = E\{[d(n) - \mathbf{w}_N^T \mathbf{x}_N(n)]^2\}, \qquad (2.2.5)$$

where $\mathbf{w}_N(n) = \mathbf{w}_N$ has been used in (2.2.4). The argument of the above expectation may be expanded as

$$[d(n) - \mathbf{w}_N^T \mathbf{x}_N(n)]^2 = d^2(n) - 2\mathbf{w}_N^T d(n)\mathbf{x}_N(n) + \mathbf{w}_N^T \mathbf{x}_N(n)\mathbf{x}_N^T(n)\mathbf{w}_N. \quad (2.2.6)$$

When the expectation operator is applied to the right-hand side of (2.2.6), the linearity of the operator allows constant terms to be factored outside the operator. Constants appear in (2.2.6) since it is assumed in the present development that the filter coefficients are fixed. Therefore, $\mathbf{w}_N$ is fixed with respect to the expectation and (2.2.6) becomes

$$\varepsilon(\mathbf{w}_N) = E\{d^2(n)\} - 2\mathbf{w}_N^T E\{d(n)\mathbf{x}_N(n)\} + \mathbf{w}_N^T E\{\mathbf{x}_N(n)\mathbf{x}_N^T(n)\}\mathbf{w}_N. \quad (2.2.7)$$

The first expectation in (2.2.7) is simply the mean square power, σ_d^2, of the desired signal. The second expectation arises quite frequently in MSE problems and is sometimes given the name cross-correlation vector. This cross-correlation vector will be denoted by $\mathbf{p}_N$, where

$$\mathbf{p}_N = E\{d(n)\mathbf{x}_N(n)\}. \qquad (2.2.8)$$

For statistically stationary signals, the cross-correlations in (2.2.8) are independent of time [13]. The $\mathbf{p}_N$ vector is given this name since it is the cross-corresponds to the physical situation of filtering the signal $x(n)$ with the fixed to compute the prediction of $d(n)$. The components of $\mathbf{p}_N$ may be easily obtained if necessary, since by expanding (2.2.8)

$$\mathbf{p}_N = \begin{bmatrix} E\{d(n)x(n)\} \\ E\{d(n)x(n-1)\} \\ \cdots \\ E\{d(n)x(n-N+1)\} \end{bmatrix} = \begin{bmatrix} \phi_{xd}(0) \\ \phi_{xd}(1) \\ \cdots \\ \phi_{xd}(N-1) \end{bmatrix}, \qquad (2.2.9)$$

where the $\phi_{xd}(m)$ in (2.2.9) are the cross-correlation coefficients, or cross-correlation lags, defined in general by [14, 16]

$$\phi_{xy}(m) = E\{x(n)y(n-m)\} = E\{y(n)x(n-m)\}. \qquad (2.2.10)$$

Very frequently in signal-processing applications, the true value of $\phi_{xy}(m)$ is unknown and instead must be estimated from the data. One quite popular method for estimating correlation values for stationary, ergodic signals is given from [14]:

$$\hat{\phi}_{xy}(m) = \frac{1}{K} \sum_{i=0}^{K-|m|-1} x(n-i)y(n-|m|-i). \qquad (2.2.11)$$

Similarly, the third expectation in (2.2.3) appears so frequently in signal processing problems that it is also given a designation. The autocorrelation matrix, $\mathbf{R}_{NN}$, is defined as

$$\mathbf{R}_{NN} = E\{\mathbf{x}_N(n)\mathbf{x}_N^T(n)\}. \qquad (2.2.12)$$

Again, for statistically stationary signals, the $\mathbf{R}_{NN}$ matrix is independent of time n. This matrix describes the statistical relations among the individual components used in making the prediction of $d(n)$. The form of the $\mathbf{R}_{NN}$ matrix elements can be seen by expanding the product in (2.2.12) and taking expectations:

$$E\{\mathbf{x}_N(n)\mathbf{x}_N^T(n)\} = E\left\{\begin{bmatrix} x(n) \\ x(n-1) \\ \cdots \\ x(n-N+1) \end{bmatrix} [x(n), x(n-1), \ldots, x(n-N+1)]\right\}.$$

(2.2.13)

The expectation of the matrix $\mathbf{x}_N(n)\mathbf{x}_N^T(n)$ is equal to the matrix of the individual expectations. Furthermore, since $x(n)$ is assumed to be stationary, then each of the elements is of the form

$$E\{x(n-m)x(n-k)\} = \phi_x(m-k) = \phi_x(k-m),$$

(2.2.14)

where $\phi_x(p)$ is the pth autocorrelation coefficient of the stationary signal $x(n)$. Quite frequently, these autocorrelation values must be estimated directly from the available data. One popular method is to use an estimator analogous to (2.2.11):

$$\hat{\phi}_x(m) = \frac{1}{K} \sum_{i=0}^{K-m-1} x(n-i)x(n-m-i).$$

(2.2.15)

This estimate is used quite frequently [9, 14], since one of the benefits of using (2.2.15) is that the resulting estimated autocorrelation function is symmetric; that is, $\hat{\phi}(m) = \hat{\phi}(-m)$. In this case, the estimated autocorrelation matrix would therefore be symmetric.

Continuing the development for the known autocorrelation case, the autocorrelation matrix is thus given by

$$\mathbf{R}_{NN} = \begin{bmatrix} \phi_x(0) & \phi_x(1) & \cdots & \phi_x(N-1) \\ \phi_x(1) & \phi_x(0) & \cdots & \phi_x(N-2) \\ \cdots & \cdots & \cdots & \cdots \\ \phi_x(N-1) & \phi_x(N-2) & \cdots & \phi_x(0) \end{bmatrix}.$$

(2.2.16)

Note that $\mathbf{R}_{NN}$ as expressed in (2.2.16) is symmetric. This property is due to selecting the data $x(n)$, $x(n-1)$, and so on, in a sequential manner back to $x(n-N+1)$. If a prediction were made using data that did not extend sequentially N samples into the past, then the $\mathbf{R}_{NN}$ matrix would not necessarily be symmetric.

Returning to (2.2.7), make the substitutions (2.2.8) and (2.2.12) to derive the form

$$\varepsilon(\mathbf{w}_N) = \sigma_d^2 - 2\mathbf{w}_N^T\mathbf{p}_N + \mathbf{w}_N^T\mathbf{R}_{NN}\mathbf{w}_N.$$

(2.2.17)

The MSE is immediately seen to be a function of the predictor filter coefficients

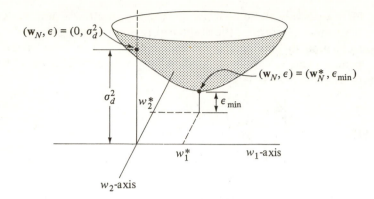

Figure 2.1   Location of the minimum of the MSE surface and its relation to $\mathbf{w}_N^*$ and $\varepsilon_{\min}$.

$\mathbf{w}_N$. If any component of $\mathbf{w}_N$ on the right-hand side is allowed to change, then the computed value of MSE on the left-hand side could also change. Since $\mathbf{w}_N$ can assume a continuum of values in the N dimension $\mathbf{w}_N$-plane, the MSE values associated with $\mathbf{w}_N$ therefore form a continuum in an $N + 1$ dimension space. For $N = 2$, this corresponds to an error "surface" in a three-dimensional space, which is commonly called the *mean square error surface*.

This situation is shown geometrically in Figure 2.1 for the case $N = 2$. The specific labels in Figure 2.1 will be developed in the next section. For now, only a qualitative description of the MSE surface is necessary. The height $\varepsilon(\mathbf{w}_N)$ corresponds to the physical situation of filtering the signal $x(n)$ with the fixed filter $\mathbf{w}_N$, from which a prediction error signal $e(n)$ with power $\varepsilon(\mathbf{w}_N)$ is generated. If the filter coefficients are changed, then it stands to reason that the power in the error signal might change. This is indicated by the changing height of the surface above the $\mathbf{w}_N$-plane as the component values of $\mathbf{w}_N$ are varied. Some filter setting $(w_1, w_2) = (w_1^*, w_2^*)$ will produce the minimum MSE $\varepsilon_{\min}$.

Example

An example from systems identification will help give the mathematical development a physical association. Consider the case in Figure 2.2(a), in which an $N = 2$ coefficient transversal filter with coefficients w_k tries to predict the output of a system h_k using the input $x(n)$ to the system. Suppose the true system h_k is actually a third-order system with impulse response $(h_0, h_1, h_2) = (0.2, 0.7, 0.2)$. Let the input $x(n)$ be a random zero mean gaussian signal as shown in Figure 2.2(b) and the output therefore is given by

$$d(n) = \sum_{k=0}^{2} h_k x(n - k), \qquad (2.2.18)$$

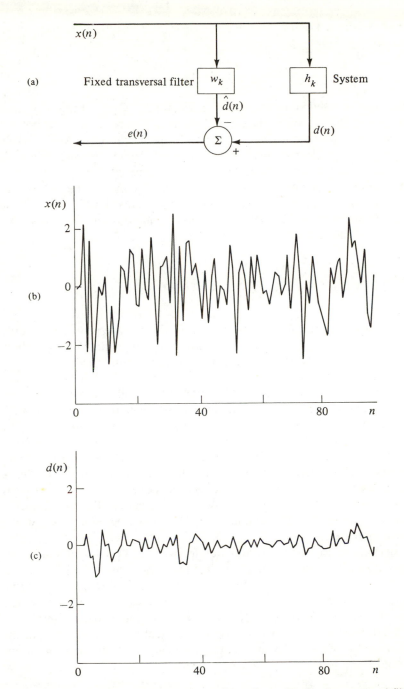

Figure 2.2   (a) Systems identification configuration using fixed transversal filter w_k. (b) Zero mean uncorrelated gaussian signal $x(n)$ used in systems identification. (c) Output of system h_k using $x(n)$ as input.

which is shown in Figure 2.2(c). The error signals resulting from trying to predict $d(n)$ by using

$$\hat{d}(n) = \sum_{k=0}^{1} w_k x(n - k) \qquad (2.2.19)$$

for three different sets of w_k are shown in Figure 2.3. Each setting of w_1 and w_2 corresponds to a point in the (w_1, w_2) plane of Figure 2.1, and the powers of the resulting error sequences correspond to the height of the MSE surface above the (w_1, w_2). Clearly, each filter setting produces an error sequence with a different power (or variance), and there will be some set of weights that will give the lowest error power. This is represented by the setting (w_1^*, w_2^*) in Figure 2.1. Rather than randomly trying to choose this "best" set of w_k, it is desired to find an analytical method for determining this set of weights. This is the topic of the next section.

2.3 Properties of the MSE Surface

One of the most important properties of the MSE surface is that it has only one extremum, and that extremum is a minimum point. A simple way to show that the MSE surface has this property is to examine geometrically what must occur at the hypothesized minimum. Examination will be done for the case for $N = 2$, but the concept is readily extended to higher dimensions. One characteristic of the surface at the minimum point is that the tangents to the surface in directions parallel to the w_1- and w_2-axes must be zero. In other words, consider the MSE surface in Figure 2.4(a) and suppose a cross-section were cut through the error surface along the line $w_1 = w_1^*$, where w_1^* is the w_1 coordinate of the hypothesized minimum. This is illustrated in Figure 2.4(b). This line is parallel to the w_2-axis, and the independent variable is now w_2 since w_1 has been fixed at $w_1 = w_1^*$. Progressing along this line by increasing w_2, the minimum of the error surface is encountered at $w_2 = w_2^*$, after which the MSE increases.

To derive a mathematical formulation for w_1^* and w_2^* a way of mathematically stating the condition described in Figure 2.4(b) is needed. Note that the curve is concave up, or that the second derivative of $\varepsilon(\mathbf{w}_N)$ with respect to w_2 (along the line $w_1 = w_1^*$) is positive. Therefore, for a minimum of the MSE surface, the first derivative of the error surface evaluated at the minimum point must be zero and the second derivative must be positive. Thus, at the minimum point of the MSE surface, the following mathematical conditions must hold:

$$\left[\frac{\partial}{\partial w_2} \varepsilon(w_1, w_2) \right]_{w_1, w_2 = w_1^*, w_2^*} = 0, \qquad (2.3.1a)$$

and

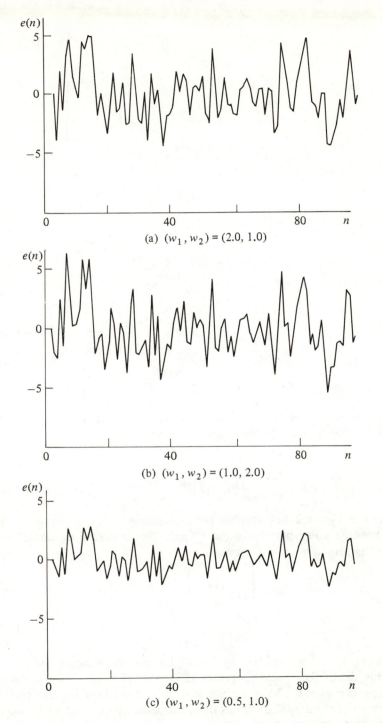

Figure 2.3 System identification error results for three arbitrarily chosen filters (w_1, w_2).

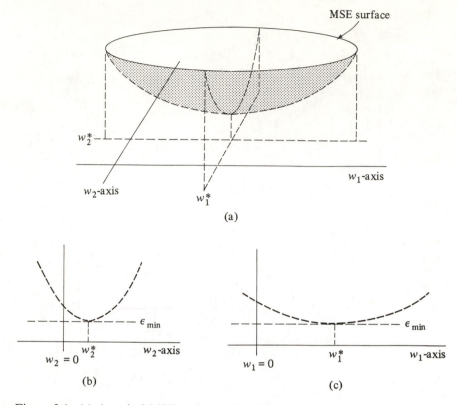

Figure 2.4   (a) A typical MSE surface.   (b) MSE cross-section with respect to w_1. (c) MSE cross-section with respect to w_2.

$$\left[\frac{\partial^2}{\partial w_2^2}\varepsilon(w_1,w_2)\right]_{w_1,w_2=w_1^*,w_2^*} > 0. \tag{2.3.1b}$$

If, as in Figure 2.4(c), another cross-section were cut through the MSE minimum on a line parallel to the w_1-axis, then a reasoning similar to the preceding produces the equations

$$\left[\frac{\partial}{\partial w_1}\varepsilon(\mathbf{w}_N)\right]_{\mathbf{w}_N=\mathbf{w}_N^*} = 0 \tag{2.3.2a}$$

$$\left[\frac{\partial^2}{\partial w_1^2}\varepsilon(\mathbf{w}_N)\right]_{\mathbf{w}_N=\mathbf{w}_N^*} > 0, \tag{2.3.2b}$$

where $(w_1,w_2)^T = \mathbf{w}_N$ and $(w_1^*,w_2^*)^T = \mathbf{w}_N^*$ has now been used for simplicity.

Thus, the mathematical properties (2.3.1) and (2.3.2) quantify the geometrical properties of the MSE surface at its minimum. To show that the MSE surface $\varepsilon(\mathbf{w}_N)$ given by (2.2.17) has these properties, expand (2.2.17) to clearly display w_1 and w_2:

$$\varepsilon(\mathbf{w}_N) = \varepsilon(w_1, w_2)$$
$$= w_1^2 \phi_x(0) + 2w_1 w_2 \phi_x(1) + w_2^2 \phi_x(0) - 2w_1 \phi_{xd}(0) - 2w_2 \phi_{xd}(1) + \sigma_d^2.$$
$$(2.3.3)$$

Then substituting (2.3.3) into (2.3.1a) and (2.3.2a) produces the following equations:

$$2w_1^* \phi_x(0) + 2w_2^* \phi_x(1) - 2\phi_{xd}(0) = 0 \qquad (2.3.4a)$$

$$2w_2^* \phi_x(0) + 2w_1^* \phi_x(1) - 2\phi_{xd}(1) = 0. \qquad (2.3.4b)$$

Note that these equations can be written as the single matrix equation

$$\mathbf{R}_{NN} \mathbf{w}_N^* = \mathbf{p}_N, \qquad (2.3.5)$$

with $\mathbf{R}_{NN}$ and $\mathbf{p}_N$ defined by (2.2.8) and (2.2.12), respectively.

The equations (2.3.5) are often called the *normal equations* and are very important in the study of linear minimum MSE filtering. If the second-order signal statistics are known, then the normal equations provide a theoretical basis for solving the optimal prediction filter $\mathbf{w}_N^*$. The importance of the result in (2.3.5) should not be underestimated because of its simplicity. If the matrix inverse $\mathbf{R}_{NN}^{-1}$ exists, then premultiplying both sides of (2.3.5) by $\mathbf{R}_{NN}^{-1}$ would produce

$$\mathbf{w}_N^* = \mathbf{R}_{NN}^{-1} \mathbf{p}_N. \qquad (2.3.6)$$

For second-order statistics representative of the vast majority of real world physical applications, the $\mathbf{R}_{NN}$ matrix will indeed be invertible. This is an especially applicable assumption if signals in broadband noise environments or signals generated by stochastic difference equation models are being considered. For further information on invertibility and the mathematical implications, the reader is referred to any of the excellent linear algebra texts [1–6]. In passing, it should be stated that all the practical signal processing applications considered in this text will produce invertible $\mathbf{R}_{NN}$ matrices. It should also be stated, however, that actual solutions $\mathbf{w}_N^*$ are rarely ever found by inverting the $\mathbf{R}_{NN}$ matrix, since more efficient methods exist for solving linear equations [1–6, 17]. Thus the result (2.3.6) is really of more theoretical than practical importance.

The problem remains concerning the uniqueness of the MSE surface minimum at location $\mathbf{w}_N^*$. Fortunately, from well-known results in linear algebra [5, 6], if the matrix $\mathbf{R}_{NN}$ is *nonsingular* (and therefore invertible), then the $\mathbf{w}_N^*$ found by (2.3.6) is the unique solution to the matrix equation (2.3.5). Again, all the practical signal processing applications considered in this text will produce nonsingular $\mathbf{R}_{NN}$ matrices.

There are two solution methods to (2.3.5), known as the method of steepest descent [15] and Durbin's algorithm [11], that lead to popular adaptive signal processing techniques. The method of steepest descent, which has led to the LMS algorithm, is an iterative technique and is the subject of Chapter 4. Durbin's algorithm, which has led to the adaptive lattice filter, constructs sequential lower order solutions to the Nth order normal equations and is investigated in Chapter 3.

The MSE surface minimum

The preceding has shown that the MSE surface has only one extremum. It remains to be shown that this is a minimum rather than a maximum. Recall that the only information given by the first derivative requirements in (2.3.4) was the location of the extremum in the $\mathbf{w}_N$-space. If the extremum is a minimum, then the cross-sectional cuts parallel to the coordinate axes should be as illustrated in Figure 2.4. Each of the curves in Figures 2.4(b) and 2.4(c) must be concave up, or, equivalently, the second derivative expressions of (2.3.1b) and (2.3.2b) must be positive. Since the first derivatives of $\varepsilon(\mathbf{w}_N)$ have already been calculated in (2.3.4), then the second derivatives required by (2.3.1b) and (2.3.2b) are

$$\left[\frac{\partial^2}{\partial w_1^2} \varepsilon(\mathbf{w}_N) \right]_{\mathbf{w}_N = \mathbf{w}_N^*} = 2\phi_x(0) \qquad (2.3.7a)$$

$$\left[\frac{\partial^2}{\partial w_2^2} \varepsilon(\mathbf{w}_N) \right]_{\mathbf{w}_N = \mathbf{w}_N^*} = 2\phi_x(0). \qquad (2.3.7b)$$

Therefore, the requirement for the error surface to be "concave up" is simply that the diagonal elements, $\phi_x(0)$, of the $\mathbf{R}_{NN}$ matrix be greater than zero. This is a reasonable assumption for practical physical signals since the definition of $\phi_x(0)$ in (2.2.14) produces

$$\phi_x(0) = E\{x^2(n - j)\} = \{x^2(n)\} = \sigma_x^2.$$

The last expressions above are valid since $x(n)$ is a stationary signal, and thus the expectation is independent of a shift in time argument. However, σ_x^2 is the mean square power of the signal $x(n)$, which must be positive for real-valued signals. Thus,

$$\phi_x(0) = \sigma_x^2 > 0,$$

and the second derivative expressions of (2.3.7) are indeed positive. Therefore, the error surface at the location of the extremum is concave up, which means that the extremum is indeed a minimum point, as hypothesized.

Additionally, the height above the $\mathbf{w}_N$-plane of this minimum point on the error surface may be found. Since this minimum height corresponds to the minimum MSE achievable, define this as the minimum MSE, $\varepsilon_{\min}$. Substituting (2.3.6) into (2.2.17) provides an expression for $\varepsilon_{\min}$:

$$\varepsilon_{\min} = \varepsilon(\mathbf{w}_N^*), \qquad (2.3.8)$$

which after some intermediate algebra produces

$$\varepsilon_{\min} = \sigma_d^2 - \mathbf{p}_N^T \mathbf{w}_N^*. \qquad (2.3.9)$$

Thus, this minimum point on the error surface is at a height $\varepsilon_{\min}$ above the w_N-plane. This physical interpretation of (2.3.9) is that if $d(n)$ and $x(n)$ were stationary signals with known cross-correlation properties, then using w_N^* as a prediction filter would produce the minimum power error sequence, which would have power $\varepsilon_{\min}$ given in (2.3.9). Note that the right-hand side of (2.3.9) neatly divides the prediction error power into two components. If there were no prediction filtering at all, then w_N would equal zero and the error power would equal σ_d^2. The optimal prediction filter w_N^* effectively removes the portions of the desired signal $d(n)$, which are correlated with the data signal $x(n)$. This reduces the error power by a factor $p_N^T w_N^*$ and is the best that any linear MSE predictor can do, provided the signals $d(n)$ and $x(n)$ are stationary. Conversely, when the w_N are chosen such that the output prediction error signal finally has mean square power equal to $\varepsilon_{\min}$, then the optimum predictor filter coefficients have been achieved, and the search for the optimum filter may therefore be terminated. This latter interpretation represents how the adaptive methods to be examined in this text approach the iterative computation of the minimum MSE filter.

Impact of signal correlation

In order for linear minimum MSE filtering to have any impact, there must be non-zero cross-correlation between the desired signal $d(n)$ and the data signal $x(n)$. An important point not immediately obvious is that neither $d(n)$ nor $x(n)$ itself need be correlated. For example, consider again a system of the type in Figure 2.1 with uncorrelated input $x(n)$, and let the system simply delay the input by two samples and attenuate it by a factor of a, giving

$$d(n) = ax(n - 2). \tag{2.3.10}$$

Then it is easy to show from (2.2.12)

$$\mathbf{R}_{NN} = \sigma_x^2 \mathbf{I}_{NN}, \tag{2.3.11}$$

where $\mathbf{I}_{NN}$ is the identity matrix. Then from (2.2.9)

$$\mathbf{p}_N = [0, 0, a\sigma_x^2, 0, \dots]^T. \tag{2.3.12}$$

The minimum MSE predictor is then given from (2.3.6) by

$$\mathbf{w}_N^* = [0, 0, a, 0, \dots]^T. \tag{2.3.13}$$

Moreover, (2.3.9) shows that the minimum MSE is given by

$$\varepsilon_{\min} = \sigma_d^2 - a^2 \sigma_x^2 = 0, \tag{2.3.14}$$

since $\sigma_d^2 = a^2 \sigma_x^2$ from (2.3.10). Thus, it is possible (theoretically, at least) to obtain zero prediction error because of the cross-correlation between $x(n)$ and $d(n)$.

However, now consider the case of linear prediction of the uncorrelated signal $x(n)$; that is, it is desired to predict $x(n)$ using past samples $x(n - 1), \ldots,$ $x(n - N)$:

$$\hat{x}(n) = \mathbf{w}_N^T \mathbf{x}_N(n - 1), \tag{2.3.15}$$

where the notation of (2.2.4c) has been used. Thus, from (2.2.12)

$$\mathbf{R}_{NN} = E\{\mathbf{x}_N(n - 1)\mathbf{x}_N^T(n - 1)\} = \sigma_x^2 \mathbf{I}_{NN}, \tag{2.3.16}$$

but since $x(n)$ is uncorrelated, (2.2.9) gives

$$\mathbf{p}_N = E\{x(n)\mathbf{x}_N(n - 1)\} = 0. \tag{2.3.17}$$

Thus, the minimum MSE filter in this case is

$$\mathbf{w}_N^* = 0. \tag{2.3.18}$$

The "optimal" filter is zero! This result is actually quite reasonable since there is no way to accurately predict the next sample in an uncorrelated signal. Thus, *any* prediction would be erroneous and only add to the error power. A zero prediction results in an MSE of σ_x^2, which is the best any filter could do for predicting an uncorrelated signal.

2.4 The Normal Equations

The normal equations (2.3.5) give the relation between the signal environment (represented by the statistical properties $\mathbf{p}_N$ and $\mathbf{R}_{NN}$) and the minimum MSE filter, $\mathbf{w}_N^*$. This is important since it guarantees that if the second-order statistics of the data are known, then the optimum MSE predictor can be computed. The normal equations provide an extremely important statement about mean square prediction and filtering, and their impact extends into such diverse fields as signal processing, speech, communications, radar, economics, and underwater acoustics, to name a few. However, while in theory the normal equations do allow a computation of the minimum MSE filter, some computational difficulties often prohibit its direct solution via matrix inversion. A gradient-based technique known as the method of steepest descent will be developed in Chapter 4, which allows these difficulties to be overcome. Then in Chapter 5, an approximation to steepest descent will lead to the popular and widely used LMS algorithm. However, the normal equations are very important in their own right, since they provide a theoretical basis for solving prediction and estimation problems via adaptive methods.

Data and prediction error orthogonality

In the previous section, the normal equations were derived from the geometrical requirements at the minimum of the MSE surface. The normal equations may also be derived in a different manner, displaying a fundamental property of minimum MSE filters known as the *orthogonality conditions*. To derive these conditions, minimize the MSE with respect to $\mathbf{w}_N^*$:

$$\frac{\partial}{\partial \mathbf{w}_N} E\{e^2(n)\} = E\left\{\frac{\partial}{\partial \mathbf{w}_N} e^2(n)\right\} = E\left\{2e(n)\frac{\partial}{\partial \mathbf{w}_N} e(n)\right\} = 0, \qquad (2.4.1)$$

where the linearity of the gradient and expectation operators has been used to interchange their order. Also from (2.2.5)

$$\frac{\partial}{\partial \mathbf{w}_N} e(n) = \frac{\partial}{\partial \mathbf{w}_N} [d(n) - \mathbf{w}_N^T \mathbf{x}_N(n)] = -\mathbf{x}_N(n). \qquad (2.4.2)$$

Substituting (2.4.2) into (2.4.1) and simplifying gives

$$E\{e(n)\mathbf{x}_N(n)\} = 0. \qquad (2.4.3)$$

Expanding (2.4.3) then produces the N simultaneous equations

$$E\{e(n)x(n - i)\} = 0, \qquad 0 \le i \le N - 1. \qquad (2.4.4)$$

This very important set of conditions is known as the *orthogonality conditions* and results directly from minimizing the MSE. The term orthogonality stems from the statistical definition of the result in (2.4.4); that is, two random sequences are defined to be orthogonal if [13]

$$E\{p(n)q(n)\} = 0. \qquad (2.4.5)$$

The orthogonality condition states that the expectation of the product of the prediction error $e(n)$ with the data used in the prediction is zero. The orthogonality condition must be maintained by the minimum MSE predictor, regardless of the manner in which the predictor coefficients are computed. One immediate implication is that if either $e(n)$ or $x(n)$ is zero mean, then the orthogonality conditions imply that the error sequence is uncorrelated with the data. For example, suppose that $E\{x(n)\} = 0$, as is true in many speech and communications applications. If $x(n)$ is a stationary signal, then $E\{x(n - i)\} = 0$ for any shift i. Multiplying this result by the expectation of the error then gives

$$E\{e(n)\}E\{(n - i)\} = 0; \qquad -\infty \le i \le \infty. \qquad (2.4.6)$$

But the orthogonality condition requires

$$E\{e(n)x(n - i)\} = 0; \qquad 0 \le i \le N - 1. \qquad (2.4.7)$$

Equating these results then provides

$$E\{e(n)x(n - i)\} = E\{e(n)\}E\{x(n - i)\},$$

which is simply a statement that $e(n)$ and $x(n - i)$ are uncorrelated. Therefore, the prediction error is uncorrelated with the data used in making the prediction. The uncorrelated and orthogonality properties will be referenced frequently in the remainder of the text, since they are central to the concept of mean square filtering, prediction, and estimation.

The normal equations are a natural consequence of the orthogonality conditions and, in fact, derive their name from this property. The geometrical property of "right angles" may be associated with the statistical property of orthogonality from (2.4.5). Two vectors in Cartesian space that are orthogonal (i.e., at right angles to each other) are said to be "normal" with respect to one another. Hence, the term *normal* to describe the set of N conditions in (2.4.7).

To derive the normal equations from the orthogonality conditions, simply expand $e(n)$ in (2.4.3) using the definition of prediction error and evaluate the resulting expression at $\mathbf{w}_N = \mathbf{w}_N^*$:

$$[E\{\mathbf{x}_N(n)[d(n) - \mathbf{x}_N^T(n)\mathbf{w}_N]\}]_{\mathbf{w}_N=\mathbf{w}_N^*} = 0. \tag{2.4.8}$$

Expanding (2.4.8) produces

$$E\{\mathbf{x}_N(n)\mathbf{x}_N^T(n)\mathbf{w}_N^*\} = E\{d(n)\mathbf{x}_N(n)\}. \tag{2.4.9}$$

But $\mathbf{w}_N^*$ is a constant vector. Therefore, (2.4.9) may be rewritten as

$$E\{\mathbf{x}_N(n)\mathbf{x}_N^T(n)\}\mathbf{w}_N^* = E\{d(n)\mathbf{x}_N(n)\}. \tag{2.4.10}$$

But the expectations in (2.4.10) have been defined from (2.2.9) and (2.2.12), and (2.4.10) becomes

$$\mathbf{R}_{NN}\mathbf{w}_N^* = \mathbf{p}_N, \tag{2.4.11}$$

which agrees with (2.3.5), which was based upon a geometrical consideration of the MSE surface.

Thus, the normal equations have been alternatively derived from the mathematical perspective of fulfilling the orthogonality conditions. The same equations were previously produced by fulfilling the geometrical requirements at the minimum of the error surface. In later chapters, both mathematical and geometrical approaches will often be used in combination to investigate the same problem. Each approach complements the other and often gives insight not immediately obvious from the consideration of only a single approach.

2.5 Further Geometrical Properties of the Error Surfaces

The search for the optimum MSE filter $\mathbf{w}_N^*$ may equivalently be considered as the search for the minimum of the MSE surface. Therefore, the orientation of the MSE surface in the $N + 1$ dimension space will greatly affect the performance of different algorithms in finding this minimum. One method of describing the orientation of this surface is to determine its principal axes, which will be shown to be related to the eigenvalues and eigenvectors of the autocorrelation matrix $\mathbf{R}_{NN}$. These and other properties of the MSE surface are developed in this section.

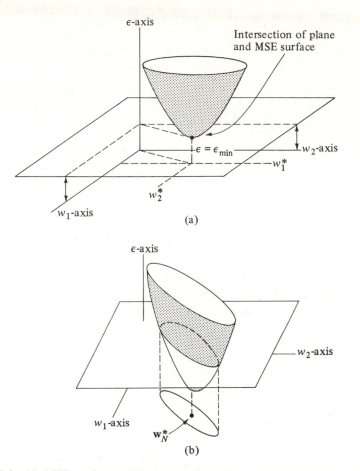

Figure 2.5   (a) MSE surface and its intersection with plane at $\varepsilon = \varepsilon_{\min}$.   (b) Inter-section of $\varepsilon = 2\varepsilon_{\min}$ plane and MSE surface, and the resulting projection onto the $\mathbf{w}_N$-plane.

Eigenvalues and eigenvectors of the autocorrelation matrix

Equation (2.2.17) provides a great deal more information about the error surface than just its height above the $\mathbf{w}_N$-plane. To see this, begin by rewriting (2.2.17) slightly as

$$\mathbf{w}_N^T \mathbf{R}_{NN} \mathbf{w}_N - 2\mathbf{p}_N^T \mathbf{w}_N - (\varepsilon - \sigma_d^2) = 0. \qquad (2.5.1)$$

Now consider the operations depicted in Figure 2.5, which displays the case for $N = 2$ coefficients. Suppose the MSE surface were intersected by a plane parallel to the $\mathbf{w}_N$-plane at a height $\varepsilon = \varepsilon_{\min}$ above the $\mathbf{w}_N$-plane. At every point of intersection of the error surface with this plane construct a projection onto

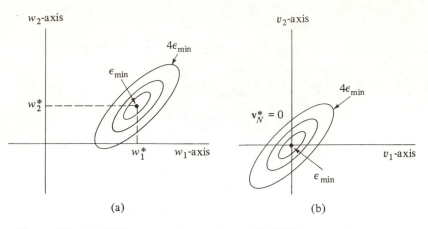

Figure 2.6   (a) MSE contours in $\mathbf{w}_N$-plane.   (b) MSE contours in $\mathbf{v}_N$-plane.

the $\mathbf{w}_N$-plane, such that the locus of points satisfying the intersection in the original $N + 1$ dimension space projects to a locus in the N dimension $\mathbf{w}_N$-plane. It is easy to see in Figure 2.5(a) that only one point on the $N + 1$ dimension error surface lies in the intersection of the error surface with the plane $\varepsilon = \varepsilon_{\min}$. Therefore, the point at "height" $\varepsilon_{\min}$ on the MSE surface corresponds to the optimal filter location (w_1^*, w_2^*) in the $\mathbf{w}_N$-plane. Another way of interpreting this is that only one set of filter coefficients, namely, $\mathbf{w}_N = \mathbf{w}_N^*$, gives the minimum MSE value $\varepsilon_{\min}$.

Now suppose it is desired to find all the filter coefficient settings $\mathbf{w}_N$ that give a mean square prediction error equal to $2\varepsilon_{\min}$. This situation is illustrated geometrically in Figure 2.5(b), in which the error surface is intersected by the plane at constant MSE value $\varepsilon = 2\varepsilon_{\min}$. The filter coefficients that would produce this MSE are therefore located on the contour in the $\mathbf{w}_N$-plane formed by projecting the intersection of the MSE surface and the plane $\varepsilon = 2\varepsilon_{\min}$ onto the $\mathbf{w}_N$-plane. A solution for the locus of these $\mathbf{w}_N$ may be found by evaluating (2.5.1) with $\varepsilon = 2\varepsilon_{\min}$ and then solving the resulting equations for the $\mathbf{w}_N$. Evaluation of (2.5.1) at $\varepsilon = 2\varepsilon_{\min}$ produces:

$$\mathbf{w}_N^T \mathbf{R}_{NN} \mathbf{w}_N - 2\mathbf{p}_N^T \mathbf{w}_N - (2\varepsilon_{\min} - \sigma_d^2) = 0. \qquad (2.5.2)$$

A common term for an MSE locus in the $\mathbf{w}_N$-plane is MSE *contour*, which will be employed in this text. A general set of MSE contours is shown in Figure 2.6(a). It will be of substantial analytical interest to solve for these MSE contours, since this will provide insight into the qualitative properties of adaptive filter behavior. It is frequently easier to solve (2.5.2) for the MSE contour by first making a translation of coordinate axes, such that the new coordinate origin is at the $\mathbf{w}_N$ location of the optimal filter $\mathbf{w}_N^*$. Translation and rotation of coordinate axes are frequently used techniques in describing the properties of lines, graphs, and higher dimension geometrical shapes

[2, 3, 5]. As evident from Figure 2.6, there is symmetry to the error contour about the point $\mathbf{w}_N = \mathbf{w}_N^*$, and exploiting this symmetry will greatly simplify the analytical procedure of solving for the required locus. Therefore, define the new "translated" filter coordinate space, $\mathbf{v}_N$, by

$$\mathbf{v}_N = \mathbf{w}_N - \mathbf{w}_N^*. \tag{2.5.3}$$

This produces a set of $\mathbf{v}_N$-plane MSE contours centered about the origin as shown in Figure 2.6(b). Note that (2.5.3) defines the translated filter coefficients as the difference between a set of general filter coefficients $\mathbf{w}_N$ and the optimal coefficients $\mathbf{w}_N^*$. Therefore, using (2.3.6) produces

$$\mathbf{w}_N = \mathbf{v}_N + \mathbf{R}_{NN}^{-1}\mathbf{p}_N. \tag{2.5.4}$$

Substituting (2.5.4) into (2.5.2) then gives after some intermediate algebra

$$\mathbf{v}_N^T \mathbf{R}_{NN} \mathbf{v}_N - 2\varepsilon_{\min} + \sigma_d^2 - \mathbf{p}_N^T \mathbf{w}_N^* = 0. \tag{2.5.5}$$

But from (2.3.9), the last two terms above are equivalent to $\varepsilon_{\min}$. Therefore, (2.5.5) may be rewritten simply as

$$\mathbf{v}_N^T \mathbf{R}_{NN} \mathbf{v}_N = \varepsilon_{\min}. \tag{2.5.6}$$

While (2.5.6) is a more compact representation than (2.5.2), there is an additional property that may be exploited to simplify the analysis even more. From Figure 2.6(b), it is seen that the MSE contours have a set of principal axes, which, in general, are not aligned with the $\mathbf{v}_N$ coordinate axes. If these principal axes could be aligned with the coordinate axes, or, equivalently, the contours rotated to produce alignment, then the mathematical form describing the aligned contours will be very simple indeed.

From results in linear algebra [4, 5], the coordinate rotation between two sets of axes $\mathbf{v}_N$ and $\mathbf{v}_N'$ is defined by the transformation

$$\mathbf{v}_N = \mathbf{A}_{NN} \mathbf{v}_N' \tag{2.5.7}$$

$$\mathbf{v}_N' = \mathbf{A}_{NN}^T \mathbf{v}_N, \tag{2.5.8}$$

where $\mathbf{A}_{NN}$ is the *orthogonal* transformation matrix relating the two set of axes. Also, in (2.5.8) the relation for orthogonal matrices that $\mathbf{A}_{NN}^T = \mathbf{A}_{NN}^{-1}$ has been used. The matrix $\mathbf{A}_{NN}$ effectively aligns the principal axes of the contours with the axes of the $\mathbf{v}_N'$ coordinate system. The result is a very simple mathematical expression for the MSE contours that clearly displays the properties of the MSE surface.

In order to determine the specific orthogonal matrix $\mathbf{A}_{NN}$, which has the desired rotation property, substitute (2.5.7) into (2.5.6), producing

$$(\mathbf{A}_{NN} \mathbf{v}_N')^T \mathbf{R}_{NN} (\mathbf{A}_{NN} \mathbf{v}_N') = \varepsilon_{\min},$$

which after some algebra gives

$$\mathbf{v}_N'^T (\mathbf{A}_{NN}^T \mathbf{R}_{NN} \mathbf{A}_{NN}) \mathbf{v}_N' = \varepsilon_{\min}. \tag{2.5.9}$$

Since the autocorrelation matrix $\mathbf{R}_{NN}$ is real and symmetric, then there exists a similarity transformation for $\mathbf{R}_{NN}$ [4,6] in terms of its eigenvalues and eigenvectors. This transformation is given by

$$\mathbf{M}_{NN}^T \mathbf{R}_{NN} \mathbf{M}_{NN} = \mathbf{\Lambda}_{NN}, \tag{2.5.10a}$$

and

$$\mathbf{R}_{NN} = \mathbf{M}_{NN} \mathbf{\Lambda}_{NN} \mathbf{M}_{NN}^T, \tag{2.5.10b}$$

where $\mathbf{M}_{NN}$ is the $N \times N$ matrix whose columns are the eigenvectors of $\mathbf{R}_{NN}$ ($\mathbf{M}_{NN}$ is sometimes called the modal matrix of $\mathbf{R}_{NN}$) and $\mathbf{\Lambda}_{NN}$ is the $N \times N$ diagonal matrix of eigenvalues (λ_i) of $\mathbf{R}_{NN}$. Thus, if $\mathbf{A}_{NN}$ in (2.5.9) is chosen as the modal matrix of $\mathbf{R}_{NN}$, then (2.5.10a) gives

$$\mathbf{v}_N'^T \mathbf{A}_{NN} \mathbf{v}_N' = \varepsilon_{\min} \tag{2.5.11}$$

Equation (2.5.11) suggests a well-known geometrical form for the MSE contours in the $\mathbf{v}_N'$-plane. This is easily seen for the simple case of $N = 2$ by expanding (2.5.11) in terms of the v_i' and eigenvalues λ_i. It is straightforward to derive the form

$$\left[\frac{v_1'}{a}\right]^2 + \left[\frac{v_2'}{b}\right]^2 = 1, \tag{2.5.12}$$

where

$$a = (\varepsilon_{\min}/\lambda_1)^{1/2} \tag{2.5.13}$$

$$b = (\varepsilon_{\min}/\lambda_2)^{1/2}. \tag{2.5.14}$$

Equation (2.5.12) is immediately recognized to be the standard form for an ellipse with a and b the semi-major and semi-minor axes, respectively. Therefore, the MSE contour in the $\mathbf{v}_N'$-plane corresponding to the MSE of $2\varepsilon_{\min}$ on the error surface is an ellipse, centered at the origin, which intersects the $\mathbf{v}_N'$ axes at $v_1' = a$ and $v_2' = b$.

Thus, it has been shown that the MSE contour $\varepsilon = 2\varepsilon_{\min}$ forms an ellipse in the $\mathbf{v}_N'$-plane. The extension to a general value of MSE, ε_c, is also straightforward and is explored in the problems at the end of the chapter. The result is that any constant MSE contour in the $\mathbf{v}_N'$-plane forms an ellipse centered at the origin, with major and minor axes aligned with the axes of the $\mathbf{v}_N'$ coordinate space. The contours intersect the $\mathbf{v}_N'$ axes at values dependent upon the eigenvalues of $\mathbf{R}_{NN}$ and the specific MSE value chosen.

These results allow a simple method for visualizing the MSE surface via its error contours. The general properties and overall "shape" of the error contours in the original $\mathbf{w}_N$-plane and the new rotated-and-translated $\mathbf{v}_N'$-plane are the same. The contours in the $\mathbf{v}_N'$-plane have only been rotated and translated to provide the much simpler mathematical form of (2.5.12), which will be very beneficial in analytical work. If analysis were done instead with the original equation (2.5.1), then the explicit form of the ellipse would have been quite difficult to determine, due to the presence of cross-products of the

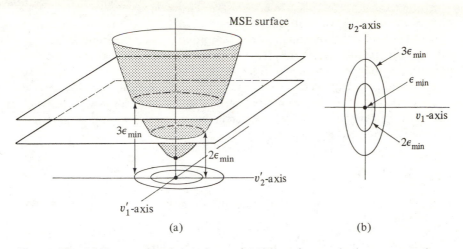

Figure 2.7 (a) Intersection in $\mathbf{v}'$-plane of MSE surface and planes at $\varepsilon = 2\varepsilon_{\min}$ and $\varepsilon = 3\varepsilon_{\min}$. (b) Projection of $2\varepsilon_{\min}$ and $3\varepsilon_{\min}$ contours in $\mathbf{v}'_N$-plane.

weights and additional linear weight terms. The translation of (2.5.3) and the rotation (2.5.8) do not alter the shape of the MSE surface. Only its orientation relative to some set of coordinate axes is changed. These transformations are very useful and quite powerful and will be used again in Chapters 4 and 5 to simplify the interpretation of adaptive filter properties.

One additional property of the MSE surface is that the entire family of MSE contours are only scaled versions of that for $\varepsilon = \varepsilon_{\min}$. To see this, consider the case of Figure 2.7(a), in which the MSE surface is intersected at heights $\varepsilon = 2\varepsilon_{\min}$ and $\varepsilon = 3\varepsilon_{\min}$. The intersections of these planes with the MSE surface result in the ellipses shown in Figure 2.7(b). If successive contours for $2\varepsilon_{\min}$ and $3\varepsilon_{\min}$ are very close together, then the "three-dimensional" MSE surface is very steep, signifying that the mean squared prediction error is very sensitive to the choice of filter coefficients. Similarly, successive contours spaced widely apart signify a mean square prediction error that is not extremely sensitive to the choice of prediction filter coefficients.

In summary, the conept of the elliptical MSE contours has been developed as a simpler interpretation of the mean square prediction error properties than the original MSE surface. All necessary information is preserved in the MSE contours; however, the dimensionality is decreased by one, which often aids in visualization.

One final property of the error surface (and, hence, the error contours) is that the orientation of the surface is entirely a function of the data statistics, and not of the filter parameters. This can be seen by recalling (2.2.17):

$$\varepsilon(\mathbf{w}_N) = \sigma_d^2 - 2\mathbf{p}_N^T \mathbf{w}_N + \mathbf{w}_N^T \mathbf{R}_{NN} \mathbf{w}_N. \tag{2.2.17}$$

Note that when $\mathbf{w}_N$ is fixed the remaining variables, $(\sigma_d^2, \mathbf{p}_N, \mathbf{R}_{NN})$ are related to the second-order statistics of the data signals. Therefore, selecting a specific filter vector $\mathbf{w}_N$ is equivalent to choosing a point in the $\mathbf{w}_N$-plane, and the height of the MSE surface above the plane at that point is then determined only by the signal correlation properties. The height of the surface above the $\mathbf{w}_N$-plane at that point may become higher or lower as the signal statistics change, which translates into an expansion or contraction of the elliptical error loci. These loci may also be rotated or translated, as has been shown. However, the basic quadratic property of (2.2.17) will be perserved, no matter what the numerical values of σ_d^2, $\mathbf{p}_N$, and $\mathbf{R}_{NN}$ may be.

PROBLEMS

1. Each set of filter coefficients $\mathbf{w}_N$ may be considered as a vector in an N dimension space.
   (a) Draw geometrical vectors corresponding to the filter vectors with coefficients given by: (i) $\mathbf{w}_N = [-2, 5]^T$, (ii) $\mathbf{w}_N = [-1, 0, 3]^T$, (iii) $\mathbf{w}_N = [0, 1, 0]^T$.
   (b) Write the convolutional filter form corresponding to (2.2.4a) for each case in 1(a).
   (c) What are the filter vectors $\mathbf{w}_N$ corresponding to the convolutional filter representations given by:
   (i) $x(n) = 3x(n-1) - 2x(n-2)$,
   (ii) $y(n) = \sum_{i=1}^{3} e^{-1}x(n-i)$,
   (iii) $z(n) = \sum_{k=1}^{6} [1 + (-1)^k]x(n-2k)$.

2. Consider a situation as shown in Figure P1 where for some reason, the processor must wait Δ samples before a prediction of $d(n)$ may be made. The notation $z^{-\Delta}$ in this figure is the Z-transform notation for this delay. This is a more general case of the situation examined in section 2.2, in which $d(n)$ was predicted based upon $x(n)$. If this modified prediction is given by

$$\hat{d}_\Delta(n) = \sum_{i=0}^{N-1} w_i x(n - \Delta - i), \tag{P.1}$$

then derive the form of the minimum MSE filter, $\mathbf{w}_N^*$, which predicts $d(n)$ based upon the delayed data signals.

3. Show that for the optimal N-length prediction filter $\mathbf{w}_N^*$, the minimum MSE may be written as

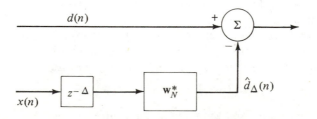

Figure P1. Optimal Δ-delayed predictor configuration.

$$\varepsilon_{min} = \sigma_d^2 - \mathbf{p}_N^T \mathbf{w}_N^*,$$

where all quantities in the above are as defined in Chapter 2.

4. Given a set of data $x(n)$ characterized by an autocorrelation function with values

$$\phi_x(0) = 1.00 \qquad \phi_x(2) = 0.40,$$
$$\phi_x(1) = 0.80 \qquad \phi_x(3) = 0.10.$$

It is desired to make a prediction of the current sample $x(n)$ based upon past samples.

(a) Find the optimum (minimum MSE) one coefficient prediction filter.

(b) Find the optimum two coefficient prediction filter.

(c) Find the optimum three coefficient prediction filter.

(d) Using the results in (a), (b), and (c), now find the minimum MSE (ε_{min}) for each filter. Concerning the complexity/performance tradeoffs, what would you say is the "best" order filter to use. Justify your answer.

5. Draw the MSE contours in the $\mathbf{v}_N'$ plane corresponding to MSEs of (i) ε_{min}, (ii) $2\varepsilon_{min}$, and (iii) $4\varepsilon_{min}$ for the eigenvalue matrices given by:

$$\text{(a) } \Lambda = \begin{bmatrix} 9 & 0 \\ 0 & 9 \end{bmatrix}, \qquad \text{(b) } \Lambda = \begin{bmatrix} 9 & 0 \\ 0 & 1 \end{bmatrix}, \qquad \text{(c) } \Lambda = \begin{bmatrix} 1/9 & 0 \\ 0 & 1 \end{bmatrix}.$$

6. Suppose it is known that a signal $x(n)$ has an autocorrelation function given by

$$\phi_x(m) = \sigma_x^2 a^{|m|}; \qquad m = 0, 1, 2, \ldots \qquad \text{(P.2)}$$

where σ_x^2 and a are fixed values. This is an example of the autocorrelation function of a *first-order Markov process*, which is a widely used statistical model for many informa- tion signals. An important characteristic of a Markov process is that the present value of the signal is dependent only upon the immediately past value $x(n-1)$.

(a) With this property in mind, derive the optimal $N = 1$ and $N = 2$ coefficient filters for predicting $x(n)$, based upon the N previous values of the signal described by (P.2). Find the corresponding ε_{min} for each case.

(b) Does the result for $N = 2$ violate your physical intuition? What property of the process causes this phenomenon? What would you expect the result to be for increasing the predictor length to $N = 3$?

7. Using the eigenvector-eigenvalue approach, draw the MSE contours corresponding ε_{min} and $4\varepsilon_{min}$ for the following cases. Assume the application is such that $d(n) = x(n)$:

$$\text{(a) } \mathbf{R}_{NN} = \begin{bmatrix} 1.0 & 0.9 \\ 0.9 & 1.0 \end{bmatrix}, \qquad \mathbf{p}_N = \begin{bmatrix} 0.9 \\ 0.7 \end{bmatrix},$$

$$\text{(b) } \mathbf{R}_{NN} = \begin{bmatrix} 1.0 & 0.5 \\ 0.5 & 1.0 \end{bmatrix}, \qquad \mathbf{p}_N = \begin{bmatrix} 0.5 \\ 0.2 \end{bmatrix}.$$

(c) The ratio of the maximum eigenvalue of a matrix to its minimum eigenvalue,

$$C = \frac{\lambda_{max}}{\lambda_{min}}$$

is called the condition number of the matrix (or sometimes called the eigenvalue spread). If C is large, then the matrix is said to be *ill-conditioned*. Relate the condition number C of the $\mathbf{R}_{NN}$ matrices in (a) and (b) above to the MSE contours and explain how the numerical property of ill conditioning affects the MSE surface.

8. Matrices that are *positive definite* are very important in linear prediction filtering. One property of positive definite matrices is that all eigenvalues are positive. Use this property to show that if the autocorrelation matrix $\mathbf{R}_{NN}$ as given by (2.2.16) is positive definite, then

$$\mathbf{z}_N^T \mathbf{R}_{NN} \mathbf{z}_N > 0$$

for any general vector $\mathbf{z}_N$.

9. The optimal filter $\mathbf{w}_N^*$ provides the lowest mean square prediction error of any linear prediction filter. One measure of comparing the performance of $\mathbf{w}_N^*$ to any other filter $\mathbf{w}_N = \mathbf{w}_N^* + \Delta\mathbf{w}_N$ is the distance measure $d(\mathbf{w}_N^*, \mathbf{w}_N)$ given by

$$\mathbf{d}(\mathbf{w}_N^*, \mathbf{w}_N) = \frac{\mathbf{a}_{N+1}^T \mathbf{R}_{N+1,N+1} \mathbf{a}_{N+1}}{\mathbf{a}_{N+1}^{*T} \mathbf{R}_{N+1,N+1} \mathbf{a}_{N+1}^*}$$

where $\mathbf{a}_{N+1}^T = [1, -\mathbf{w}_N^T]$, $\mathbf{a}_{N+1}^{*T} = [1, -\mathbf{w}_N^{*T}]$, and $\mathbf{R}_{N+1,N+1}$ is the $(N + 1) \times (N + 1)$ autocorrelation matrix.
   (a) Prove that the denominator of $d(\mathbf{w}_N^*, \mathbf{w}_N)$ is indeed $\varepsilon_{\min} = \varepsilon(\mathbf{w}_N^*)$.
   (b) Prove that $d(\mathbf{w}_N^*, \mathbf{w}_N) > 1$ for any $\mathbf{w}_N \neq \mathbf{w}_N^*$.

10. A stationary process autocorrelation matrix must be positive definite. What are the bounds on r_1 and r_2 for the matrix below to be an autocorrelation matrix?

$$\mathbf{R} = \begin{bmatrix} r_1 & r_2 \\ r_2 & r_1 \end{bmatrix}.$$

Draw this region in the (r_1, r_2) plane.

11. What are the normalized (i.e., unit-length) eigenvectors of *any* 2×2 stationary process autocorrelation matrix?

12. If an autocorrelation matrix has eigenvalues $\lambda_1 = 1.2$ and $\lambda_2 = 0.8$, what is the autocorrelation matrix?

REFERENCES

1. H. Anton, *Elementary Linear Algebra*, John Wiley & Sons, New York, 1973.
2. G. Strang, *Linear Algebra and Its Applications*, Academic Press, New York, 1976.
3. J.T. Moore, *Elementary Linear Algebra and Matrix Algebra: The Viewpoint of Geometry*, McGraw-Hill, New York, 1972.
4. B. Noble and J.W. Daniel, *Applied Linear Algebra*, Prentice-Hall, Englewood Cliffs, NJ, 1977.
5. B. Noble, *Applied Linear Algebra*, Prentice-Hall, Englewood Cliffs, NJ, 1969.
6. R. Bellman, *Introduction to Matrix Analysis*, McGraw-Hill, New York, 1960.
7. D.K. Fadeev and V.N. Fadeeva, *Computational Methods of Linear Algebra*, Freeman, San Francisco, 1963.

8. J. Makhoul, "Linear Prediction: A Tutorial," *Proceeding of the IEEE*, vol. 63, pp. 561–580, April 1975.

9. J.D. Markel and A.H. Gray, *Linear Prediction of Speech*, Springer-Verlag, New York, 1975.

10. J.D. Markel and A.H. Gray, "On Autocorrelation Equations as Applied to Speech Analysis," *IEEE Trans. Audio and Electroacoustics*, vol. AU-21, pp. 69–79, February 1973.

11. L.R. Rabiner and R.W. Schafer, *Digital Processing of Speech Signals*, Prentice-Hall, Englewood Cliffs, NJ, 1978.

12. N.S. Jayant and P. Noll, *Digital Coding of Waveforms*, Prentice-Hall, Englewood Cliffs, NJ, 1984.

13. A. Papoulis, *Probability, Random Variables, and Stochastic Processes*, McGraw-Hill, New York, 1984.

14. A.V. Oppenheim and R.W. Schafer, *Digital Signal Processing*, Prentice-Hall, Englewood Cliffs, NJ, 1975.

15. D.J. Wilde, *Optimal Seeking Methods*, Prentice-Hall, Englewood Cliffs, NJ, 1964.

16. M. Schwartz and L. Shaw, *Signal Processing: Discrete Spectral Analysis, Detection, and Estimation*, McGraw-Hill, New York, 1975.

17. G.H. Golub and C.F. Van Loan, *Matrix Computations*, Johns Hopkins University Press, Baltimore, MD, 1983.

CHAPTER 3

Linear Prediction and the Lattice Structure

3.1 Introduction

Chapter 2 presented a derivation of the normal equations from the standpoint of minimizing the mean square prediction error. However, in that chapter no specific method was presented for solving the normal equation (2.3.5)

$$\mathbf{R}_{NN}\mathbf{w}_N^* = \mathbf{p}_N \tag{2.3.5}$$

for the desired linear prediction filter $\mathbf{w}_N^*$. The only topic of interest then was the nonsingularity of the autocorrelation matrix $\mathbf{R}_{NN}$. If $\mathbf{R}_{NN}$ is nonsingular, then the inverse $\mathbf{R}_{NN}^{-1}$ exists, and the theoretical solution is given by

$$\mathbf{w}_N^* = \mathbf{R}_{NN}^{-1}\mathbf{p}_N. \tag{2.3.6}$$

The solution given by (2.3.6) is really of more theoretical than practical interest, however. There are two specific methods of solving (2.3.5) for the prediction coefficients that have direct application to actual systems implementations of adaptive filters. Moreover, each of these methods leads to a practical implementation of solving for the prediction coefficients directly from the acquired data. The first of these, Durbin's algorithm, is developed in this chapter and leads to the gradient lattice adaptive filter discussed in Chapter 7. The second technique, known as the method of steepest descent, is the topic of Chapter 4, and it leads to least mean squares (LMS) adaptive algorithm. A discussion of these adaptive methods is deferred until those later chapters. The present chapter continues with Section 3.2 by deriving Durbin's algorithm directly from the matrix equation (2.3.5). It will be seen that if $\mathbf{R}_{NN}$ has the specific form called the *Toeplitz* form, then Durbin's algorithm provides a very efficient method for solving the resulting system of linear equation (2.3.5) for $\mathbf{w}_N^*$. Furthermore, Durbin's method provides a stage-by-stage, or order-recursive method, in which a solution for the optimal pth order predictor may be computed from a knowledge of the optimal $(p - 1)$st order predictor.

The most important benefit of investigating Durbin's algorithm is that it leads directly to a stable computational linear filter structure known as the lattice filter, which is the subject of Section 3.3. The lattice has advantages over the transversal filter form of the linear predictor, especially when used in the source modeling application of Section 1.2. Durbin's algorithm computes a set of reflection coefficients, k_p, where $1 \leq p \leq N$, which will be seen to be used directly in the lattice structure. Moreover, if the linear prediction filter used in the synthesis filter (See section 1.2) in implemented in the lattice form with k_p computed by Durbin's algorithm, then the resulting synthesis filter is guaranteed to be stable [5] for infinite precision arithmetic. Since stable synthesis filters are required in waveform reconstruction systems (such as linear prediction coding [11, 12] and multipulse coding [13] for speech), Durbin's algorithm using the lattice is extremely valuable in practical applications.

3.2 Durbin's Algorithm

This section develops an efficient order-recursive method for solving the normal equations of (2.3.5), based upon a property known as the Toeplitz structure of the autocorrelation matrix. This requirement is often met in practice, especially in those applications in which the linear minimum mean square error (MSE) filter is used in a linear prediction mode suggested by the source modeling application of Figure 1.2. Originally derived by Levinson [1] and later modified by Durbin [2], this method is sometimes called the Levinson–Durbin algorithm. The method discussed in this chapter follows that approached by Durbin, and for simplicity will be referred to as Durbin's algorithm. Other approaches to this derivation may be found in Proakis [3] and Atal and Hanauer [4]. One application of Durbin's algorithm might indeed be the computation of the optimum transversal filter coefficients. However, a set of alternative parameters called the *reflection coefficients* result as a natural consequence of Durbin's algorithm. These reflection coefficients may be used in an alternative linear prediction filter structure known as the lattice, which, in some applications, has distinct advantages over the transversal form. Moreover, the reflection coefficients provide a very convenient parameterization of many naturally produced information signals, such as speech [8], electromagnetic signals [14], and seismic signals for geophysical exploration [15]. For example, the vocal tract of different speakers might be classified by distinguishing sets of reflection coefficients, thus forming a basis for speaker recognition. Another example is that seismic reflections from differing material in the earth can be classified by different reflection coefficients, which allows the determination of different classes of earth strata.

To develop Durbin's algorithm, recall from previous work that the optimal (minimum MSE) N coefficient predictor is defined by the normal equations

$$\mathbf{R}_{NN}\mathbf{w}_N = \mathbf{p}_N, \tag{3.2.1}$$

where the asterisk on the optimal predictor has been surpressed for simplicity. The strategy of Durbin's method is to assume the optimal $(N - 1)$st order filter has previously been computed, and then to calculate the optimal Nth order filter based on this assumption. With no loss of generality, the optimal $(N - 2)$nd order filter may then be assumed to have been previously computed and the $(N - 1)$st order filter to have been calculated based on this new assumption. This process of assumption is continued through lower orders of the filter until arriving at the simple consequence of needing only the optimal one coefficient predictor, which will be seen to involve only a simple division. This operation is done easily, from which the optimal two coefficient filter may then be computed. The calculation of filters of order 2 through N then proceeds in a forward manner, with each succeeding order using the results of the previous order. This method will be illustrated in detail in this section.

A critical assumption employed in Durbin's algorithm concerns the structure of the autocorrelation matrix in (3.2.1). Durbin's method assumes a Toeplitz structure for $\mathbf{R}_{NN}$, meaning that all elements along any diagonal of $\mathbf{R}_{NN}$ are equal. However, this constraint is not so restrictive as might first appear. Recall from Chapter 2 that the elements of the $\mathbf{R}_{NN}$ matrix are specific values of the autocorrelation function $\phi_x(m)$. Using a filter in the linear prediction mode on a stationary signal $x(n)$ indeed causes the autocorrelation matrix to have the Toeplitz property. If, however, the problem also requires esimating the $\phi_x(m)$ directly from the data, then some caution must be used in selecting the estimator such that the Toeplitz property is preserved. For instance, the covariance method [10] does not produce a Toeplitz autocorrelation matrix. However, the popular method known as the autocorrelation method for estimating the $\phi_x(m)$ does possess the necessary symmetry to create a Toeplitz autocorrelation matrix estimate, $\hat{\mathbf{R}}_{NN}$, and will be discussed in detail in this chapter. The autocorrelation method produces estimates of the $\phi_x(m)$ given as in (2.2.15), repeated here:

$$\hat{\phi}_x(m) = \frac{1}{K} \sum_{n=0}^{K-m-1} x(n)x(n - m), \tag{3.2.2}$$

where K is the total number of samples used in the estimate.

Given the Toeplitz structure of $\mathbf{R}_{NN}$, Durbin's algorithm may be derived as follows. Suppose $1 \le p \le N$ and the $(p - 1)$st order predictor is known. Therefore, the normal equations for the $(p - 1)$st order predictor are

$$\mathbf{R}_{p-1,p-1}\mathbf{w}_{p-1} = \mathbf{P}_{p-1}, \tag{3.2.3}$$

where

$$\mathbf{R}_{p-1,p-1} = \begin{bmatrix} \phi_x(0) & \phi_x(1) & \cdots & \phi_x(p-2) \\ \phi_x(1) & \phi_x(0) & & \phi_x(p-3) \\ \cdots & & \cdots & \cdots \\ \phi_x(p-2) & \phi_x(p-3) & \cdots & \phi_x(0) \end{bmatrix},$$

(3.2.4)

$$\mathbf{p}_{p-1} = \begin{bmatrix} \phi_x(1) \\ \phi_x(2) \\ \cdots \\ \phi_x(p-1) \end{bmatrix},$$

and $\mathbf{w}_{p-1}$ is the $(p-1)$st order predictor with coefficients given by

$$\mathbf{w}_{p-1} = [w_1^{(p-1)}, w_2^{(p-1)}, \ldots, w_{p-1}^{(p-1)}]^T.$$

(3.2.5)

The parenthetical superscript in (3.2.5) signifies that the scalar coefficients are elements of the $(p-1)$st order vector. The subscripts on the scalars denote the specific coefficient. For simplicity, the superscript is omitted on the vector $\mathbf{w}_{p-1}$. For reasons that will become evident soon, define the reversed-element vectors

$$\bar{\mathbf{w}}_{p-1} = [w_{p-1}^{(p-1)}, w_{p-2}^{(p-1)}, \ldots, w_1^{(p-1)}]^T$$

(3.2.6a)

$$\bar{\mathbf{p}}_{p-1} = [\phi_x(p-1), \phi_x(p-2), \ldots, \phi_x(2), \phi_x(1)]^T.$$

(3.2.6b)

Note that the notation in (3.2.6) allows the normal equations (3.2.3) to be written using the reversed notation.

$$\mathbf{R}_{p-1,p-1}\bar{\mathbf{w}}_{p-1} = \bar{\mathbf{p}}_{p-1}.$$

(3.2.7)

Using (3.2.3), the solution for the $(p-1)$st order predictor may be written as

$$\mathbf{w}_{p-1} = \mathbf{R}_{p-1,p-1}^{-1}\mathbf{p}_{p-1},$$

(3.2.8a)

or using (3.2.7)

$$\bar{\mathbf{w}}_{p-1} = \mathbf{R}_{p-1,p-1}^{-1}\bar{\mathbf{p}}_{p-1}.$$

(3.2.8b)

These expressions will be used directly.

The strategy is to derive a method whereby the optimal pth order predictor may be written as a function of the $(p-1)$st order predictor, which is assumed to be known. To do this, notice from (3.2.1), (3.2.4), and (3.2.6) that the pth order normal equations may be written in partitioned matrix form as

$$\begin{bmatrix} \mathbf{R}_{p-1,p-1} & \vdots & \bar{\mathbf{p}}_{p-1} \\ \cdots & \vdots & \cdots \\ \bar{\mathbf{p}}_{p-1}^T & \vdots & \phi_x(0) \end{bmatrix} \begin{bmatrix} \mathbf{w}_{p-1}^{(p)} \\ \cdots \\ k_p \end{bmatrix} = \begin{bmatrix} \mathbf{p}_{p-1} \\ \cdots \\ \phi_x(p) \end{bmatrix},$$

(3.2.9)

where

$$k_p = w_p^{(p)}$$

(3.2.10)

is called the pth *reflection coefficient*. The notation on the vector $\mathbf{w}_{p-1}^{(p)}$ signifies the $p - 1$ length vector, consisting of the first $p - 1$ coefficients (subscript denotes vector length) of the pth order solution (superscript denotes filter order). The term "reflection coefficient" may be attributed to early work in acoustic phenomena [5] in which the values of k_p could be related to the values of acoustical energy propagating in a vocal tract model.

The determination of the reflection coefficient is the most important part of this approach to linear prediction, and its computation provides the main advantage of using Durbin's algorithm. Note that at the pth iteration of the algorithm, k_p and $\mathbf{w}_{p-1}^{(p)}$ completely specify the pth order optimal linear predictor:

$$\mathbf{w}_p = \left[-\frac{\mathbf{w}_{p-1}^{(p)}}{k_p} - \right].$$

Hence, solutions for k_p and $\mathbf{w}_{p-1}^{(p)}$ must be obtained.

To solve for k_p and the vector $\mathbf{w}_{p-1}^{(p)}$, use the partitioned matrix form of (3.2.9) to write the two equations

$$\mathbf{R}_{p-1,p-1} \mathbf{w}_{p-1}^{(p)} + k_p \bar{\mathbf{p}}_{p-1} = \mathbf{p}_{p-1} \qquad (3.2.11a)$$

$$\bar{\mathbf{p}}_{p-1}^T \mathbf{w}_{p-1}^{(p)} + k_p \phi_x(0) = \phi_x(p). \qquad (3.2.11b)$$

Notice that (3.2.11a) is a $p - 1$ dimension vector equation and (3.2.11b) is a scalar equation. Taken together, (3.2.11a) and (3.2.11b) represent p equations in the p unknown $\mathbf{w}_{p-1}^{(p)}$ and k_p.

To solve these equations for k_p and $\mathbf{w}_{p-1}^{(p)}$, first premultiply both sides of (3.2.11a) by $\mathbf{R}_{p-1,p-1}^{-1}$, giving

$$\mathbf{w}_{p-1}^{(p)} = \mathbf{w}_{p-1} - k_p \bar{\mathbf{w}}_{p-1}, \qquad (3.2.12)$$

where (3.2.8) has been used to simplify the resulting equations. The result of (3.2.12) states that the first $p - 1$ coefficients of the pth order optimal predictor are simply a linear combination of the elements in the $(p - 1)$st order optimal predictor. The scaling factor controlling the amount of coefficient change in updating from the $(p - 1)$st order solution to the pth order solution is the reflection coefficient, k_p, which at this point is still unknown. Equation (3.2.12) explicitly displays the importance of the reflection coefficient in linear prediction. To formulate an expression for k_p, substitute (3.2.12) into (3.2.11b):

$$\bar{\mathbf{p}}_{p-1}^T [\mathbf{w}_{p-1} - k_p \bar{\mathbf{w}}_{p-1}] + k_p \phi_x(0) = \phi_x(p). \qquad (3.2.13)$$

Note that all of the parameters in (3.2.13) except k_p are assumed known at this step. Therefore, solving for k_p produces

$$k_p [\phi_x(0) - \bar{\mathbf{p}}_{p-1}^T \mathbf{w}_{p-1}] = \phi_x(p) - \bar{\mathbf{p}}_{p-1}^T \mathbf{w}_{p-1}, \qquad (3.2.14)$$

where the identity

$$\bar{\mathbf{p}}_{p-1}^T \bar{\mathbf{w}}_{p-1} = \mathbf{p}_{p-1}^T \mathbf{w}_{p-1}$$

has been used. However, the first term in brackets in (3.2.14) may be shown from (2.3.9) to be the mean square prediction error, ε_{p-1}, incurred from using the $(p-1)$st order minimum MSE forward prediction filter. Therefore, (3.2.14) may be written as

$$k_p = \frac{1}{\varepsilon_{p-1}}[\phi_x(p) - \mathbf{p}_{p-1}^T\bar{\mathbf{w}}_{p-1}]. \tag{3.2.15}$$

Although ε_p could be computed for each new order p using (2.3.9), there is a simpler recursion that may be derived. Rewrite (2.3.9) incorporating (3.2.12) and (3.2.10);

$$\begin{aligned} \varepsilon_p &= \phi_x(0) - \mathbf{p}_{p-1}^T\mathbf{w}_{p-1}^{(p)} - \phi_x(p)w_p^{(p)} \\ &= \phi_x(0) - \mathbf{p}_{p-1}^T[\mathbf{w}_{p-1} - k_p\bar{\mathbf{w}}_{p-1}] - k_p\phi_x(p) \\ &= [\phi_x(0) - \mathbf{p}_{p-1}^T\mathbf{w}_{p-1}] - k_p[\phi_x(p) - \mathbf{p}_{p-1}^T\bar{\mathbf{w}}_{p-1}]. \end{aligned} \tag{3.2.16}$$

But the first term in brackets in (3.2.16) is simply ε_{p-1}, the mean square prediction error using the optimal $(p-1)$st order predictor; the second bracketed term is $k_p\varepsilon_{p-1}$, as seen from (3.2.11). Therefore, (3.2.16) becomes

$$\varepsilon_p = \varepsilon_{p-1} - k_p\varepsilon_{p-1}k_p,$$

or simply

$$\varepsilon_p = (1 - k_p^2)\varepsilon_{p-1}. \tag{3.2.17}$$

At this, point all quantities needed to solve the normal equations (2.3.5) have been computed. For clarity, the required steps in Durbin's algorithm are summarized in Table 3.1.

Table 3.1 Durbin's Algorithm for Linear Prediction Coefficients

---

Initialize:

$$\varepsilon_0 = \phi_x(0) \tag{3.2.18a}$$

For $1 \le p \le N$:

$$k_p = \frac{1}{\varepsilon_{p-1}}\left[\phi_x(p) - \sum_{j=1}^{p-1} w_j^{(p-1)}\phi_x(p-j)\right] \tag{3.2.18b}$$

$$w_p^{(p)} = k_p. \tag{3.2.18c}$$

For $1 \le j \le p-1$:

$$w_j^{(p)} = w_j^{(p-1)} - k_p w_{p-j}^{(p-1)} \tag{3.2.18d}$$

$$\varepsilon_p = (1 - k_p^2)\varepsilon_{p-1}. \tag{3.2.18e}$$

At the end of iteration $p = N$, the optimum linear prediction filter is given by

$$w_j^* = w_j^{(N)}, \qquad 1 \le j \le N. \tag{3.2.18f}$$

3.3 Lattice Derivation

The lattice filter for linear prediction is an alternative to the transversal filter that directly produced the linear prediction filter $\mathbf{w}_N$. However, there are serveral benefits to the lattice structure, such as low roundoff noise in fixed wordlength implementations [6] and relative insensitivity to quantization noise [7]. Other derivations similar to the one that follows are contained in Prokis [3] and Rabiner and Schafer [8], and the reader is referred to those excellent texts for comparison and contrast.

Recall from Section 3.2 that in the current application, the linear prediction filter is used to predict the current signal value $x(n)$ based upon a knowledge of the sequentially occurring $x(n-1), \ldots, x(n-N)$. Thus, the prediction, $\hat{x}(n)$, may be written as

$$\hat{x}(n) = \mathbf{x}_N^T(n)\mathbf{w}_N^* = \sum_{i=1}^{N} w_i^{(N)} x(n-i). \tag{3.3.1}$$

The (N) superscript signifies that the coefficients are those of the optimal Nth order filter. Defining the prediction in this manner gives the following form for $e_N^f(n)$, the Nth order prediction error:

$$e_N^f(n) = x(n) - \sum_{i=1}^{N} w_i^{(N)} x(n-i). \tag{3.3.2}$$

The error, $e_N^f(n)$, will be called the *forward* prediction error (FPE) (denoted by the f superscript), since the predicted value $\hat{x}(n)$ is of a sample $x(n)$ that is forward in time with respect to the data being used to compute the prediction. Notice that (3.3.2) defines a filter with scalar input $x(n)$, and with a scalar output that is the error $e_N^f(n)$. This filter may thus be considered as the Nth order error filter. That is, the input to the filter is $x(n)$ and the output of the filter is only that portion of the $x(n)$ that is unpredictable using an Nth order predictor (i.e., the prediction error). The set of prediction coefficients in (3.3.2) that produces $e_N^f(n)$ is therefore called the FPE filter.

Since (3.3.2) relates a discrete filter input and output, it is quite natural to consider the z-domain transfer function for the forward error filter. Taking z-transforms [8,9] of both sides of (3.3.2) produces

$$E_N^f(z) = \left[1 - \sum_{i=1}^{N} w_i^{(N)} z^{-i} \right] X(z), \tag{3.3.3}$$

from which the Nth order FPE filter transfer function is given by

$$G_N^f(z) = \frac{E_N^f(z)}{X(z)} = 1 - \sum_{i=1}^{N} w_i^{(N)} z^{-i}. \tag{3.3.4}$$

Since (3.3.4) was derived for any general filter order N, there is no loss of generality if N is replaced by any integer p, where $1 \le p \le N$. Correspondingly, the transfer function for the pth order FPE filter is given by

$$G_p^f(z) = 1 - \sum_{j=1}^{p} w_j^{(p)} z^{-j}. \qquad (3.3.5)$$

Recall from (3.2.12) that the first $p - 1$ elements of the pth order optimal filter (denoted as the vector $\mathbf{w}_{p-1}^{(p)}$, signifying filter order (p) and vector length, $p - 1$) can be written in terms of the previously computed prediction vector $\mathbf{w}_{p-1}$. This was evidenced by (3.2.12), repeated here,

$$\mathbf{w}_{p-1}^{(p)} = \mathbf{w}_{p-1} - k_p \bar{\mathbf{w}}_{p-1}. \qquad (3.2.12)$$

Using the definition of $\bar{\mathbf{w}}_{p-1}$ from (3.2.6a) in (3.2.12) then gives the scalar recursions that are equivalent to the vector recursion in (3.2.12):

$$w_j^{(p)} = w_j^{(p-1)} - k_p w_{p-j}^{(p-1)}, \qquad 1 \le j \le p - 1. \qquad (3.3.6)$$

The current objective is to derive a recursion for $G_p^f(z)$ in terms of $G_{p-1}^f(z)$. That is, it is desired to derive a function for the pth order FPE filter transfer function in terms of the $(p - 1)$st order FPE filter transfer function. This is done by substituting (3.3.6) into (3.3.5) and recognizing that the $(p - 1)$st order FPE transfer function is given by

$$G_{p-1}^f(z) = 1 - \sum_{j=1}^{p-1} w_j^{(p-1)} z^{-j}. \qquad (3.3.7)$$

These operations produce

$$G_p^f(z) = G_{p-1}^f(z) - k_p \left[z^{-p} - \sum_{j=1}^{p-1} w_{p-j}^{(p-1)} z^{-j} \right]. \qquad (3.3.8)$$

This recursion for the pth order FPE transfer function in (3.3.8) may be simplified by considering the bracketed term on the right-hand side. This term introduces the concept of the *backward* prediction error (BPE) filter, which will be very useful in the remainder of the text. The important point, however, is to realize that the concept of BPE results entirely from using Durbin's algorithm, and is simply the term chosen to describe the operations in the brackets on the right-hand side of (3.3.8). To see that the term in brackets may be considered as due to a backward predictor, expand the $(p - 1)$st order FPE transfer function of (3.3.7):

$$\check{G}_{p-1}^f(z) = 1 - w_1^{(p-1)} z^{-1} - w_2^{(p-1)} z^{-2} - \cdots - w_{p-1}^{(p-1)} z^{-(p-1)}. \qquad (3.3.9)$$

Since $G_{p-1}^f(z)$ corresponds to the z-transform of the FPE filter, the FPE filter coefficients are the coefficients multiplying the z^{-j} terms in (3.3.9). Define this FPE filter as the vector $\mathbf{g}_{p-1}^f$, where

$$\mathbf{g}_{p-1}^f = [1, -w_1^{(p-1)}, -w_2^{(p-1)}, \dots, -w_{p-1}^{(p-1)}]^T. \qquad (3.3.10)$$

For example, in forming the prediction $\hat{x}(n)$ the "1" coefficient in $\mathbf{g}_{p-1}^f$ multiplies $x(n)$, "$-w_1^{(p-1)}$" multiplies $x(n - 1)$, and so on, until the last coefficient "$- w_{p-1}^{(p-1)}$" multiplies $x(n - p + 1)$. For reasons that will become quite clear directly, define the "backward" filter $\mathbf{g}_{p-1}^b$ as the vector of coefficients that is the time-reversed replica of (3.3.10):

$$\mathbf{g}_{p-1}^{b} = [-w_{p-1}^{(p-1)}, -w_{p-2}^{(p-1)}, \ldots, -w_{1}^{(p-1)}, 1]^{T}. \qquad (3.3.11)$$

The name backward filter is now appropriate, since the inner product of $\mathbf{g}_{p-1}^{b}$ with the p length vector of data $[x(n), x(n-1), \ldots, x(n-p+1)]^{T}$ is given by

$$x(n-p+1) - w_{1}^{(p-1)}x(n-p+2) - \cdots - w_{p-2}^{(p-1)}x(n-1) - w_{p-1}^{(p-1)}x(n)$$
$$= x(n-p+1) - \hat{x}(n-p+1),$$

where $\hat{x}(n-p+1)$ is the prediction of the signal value $p-1$ samples backward in time.

The BPE filter transfer function is therefore given by the z-transform of the filter vector in (3.3.11):

$$G_{p-1}^{b}(z) = -w_{p-1}^{(p-1)} - w_{2}^{(p-1)}z^{-1} - \cdots - w_{1}^{(p-1)}z^{-(p-2)} + z^{-(p-1)}. \qquad (3.3.12)$$

To see where this development is valuable, examine the function $G_{p-1}^{f}(z^{-1})$, created by replacing the variable z in (3.3.9) with z^{-1}:

$$G_{p-1}^{f}(z^{-1}) = 1 - w_{1}^{(p-1)}z - w_{2}^{(p-1)}z^{2} - \cdots - w_{p-1}^{(p-1)}z^{p-1}. \qquad (3.3.13)$$

Multiplying both sides of (3.3.13) by z^{-p}, rearranging, and using (3.3.12) then provides

$$z^{-p}G_{p-1}^{f}(z^{-1}) = z^{-1}[-w_{p-1}^{(p-1)} - \cdots - w_{1}^{(p-1)}z^{-(p-2)} + z^{-(p-1)}]$$
$$= z^{-1}G_{p-1}^{b}(z). \qquad (3.3.14)$$

Therefore, substituting (3.3.14) into (3.3.8) provides the dersired recursion

$$G_{p}^{f}(z) = G_{p-1}^{f}(z) - k_{p}z^{-1}G_{p-1}^{b}(z). \qquad (3.3.15)$$

Equation (3.3.15) is the z-domain recursion for the FPE filter transfer function, which is seen to be a function of both the $(p-1)$st order FPE and BPE transfer functions. Notice from (3.3.15) that a knowledge of the $(p-1)$st order BPE filter is needed to compute the pth order FPE transfer function. This recursion is easily derived by using (3.3.14) with $p-1$ replaced by p and substituting the result into (3.3.15) to obtain

$$G_{p}^{b}(z) = z^{-1}G_{p-1}^{b}(z) - k_{p}G_{p-1}^{f}(z). \qquad (3.3.16)$$

Taken as a pair, (3.3.15) and (3.3.16) define the complete transfer function of the pth order predictor. The lattice filter structure follows directly from examining the time series of the FPE and BPE filter outputs. Equation (3.3.4) was derived specifically for the Nth order filter, but is a completely general result as holds for any order $1 \le p \le N$. Therefore, if the time signal $x(n)$ is input to the FPE and BPE filter, then the z-transforms of the pth order filter outputs are

$$E_{p}^{f}(z) = X(z)G_{p}^{f}(z) \qquad (3.3.17)$$

$$E_{p}^{b}(z) = X(z)G_{p}^{b}(z), \qquad (3.3.18)$$

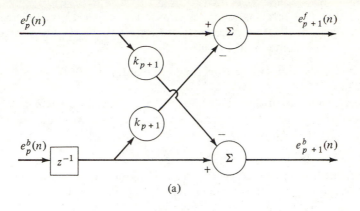

(a)

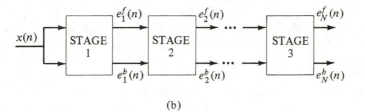

(b)

Figure 3.1 (a) Single stage of the linear prediction lattice filter. (b) Cascade of stages of form Nth order linear prediction filter.

where $X(z)$ is the z-transform of the signal $x(n)$. A simple substitution of (3.3.15) into (3.3.17) and (3.3.16) into (3.3.18) then gives

$$E_p^f(z) = X(z)G_{p-1}^f(z) - k_p z^{-1} X(z)G_{p-1}^b(z) \qquad (3.3.19)$$

$$E_p^b(z) = z^{-1} X(z)G_{p-1}^b(z) - k_p X(z)G_{p-1}^f(z). \qquad (3.3.20)$$

Simplifying the above using (3.3.17) and (3.3.18) then produces the desired recursions for the pth order error outputs:

$$E_p^f(z) = E_{p-1}^f(z) - k_p z^{-1} E_{p-1}^b(z) \qquad (3.3.21)$$

$$E_p^b(z) = z^{-1} E_{p-1}^b(z) - k_p E_{p-1}^f(z). \qquad (3.3.22)$$

The corresponding time domain expressions then follow directly from (3.3.21) and (3.3.22) by taking the inverse z-tranforms:

$$e_p^f(n) = e_{p-1}^f(n) - k_p e_{p-1}^b(n-1) \qquad (3.3.23)$$

$$e_p^b(n) = e_{p-1}^b(n-1) - k_p e_{p-1}^f(n). \qquad (3.3.24)$$

A block diagram of the time domain lattice equations is shown in Figure 3.1, which illustrates how the lattice structure computes the forward and backward errors.

The lattice structure has been widely used for implementing linear prediction filters on actual data. In summary, the lattice allows the linear prediction filter to be implemented in the following steps:

(1) Estimate the autocorrelation function values $\hat{\phi}_x(m)$, $0 \leq m \leq N$, using (3.2.2).
(2) Use Durbin's algorithm (Table 3.1) to solve for the k_p, $1 \leq p \leq N$.
(3) Use the k_p in the lattice structure (Figure 3.1), with the data signal $x(n) = e_0^f(n) = e_0^b(n)$ as input.

When used in this manner, the forward error output of the Nth stage, $e_N^f(n)$, is equivalent to the forward error out of the N coefficient transversal linear predictor. Since $e_N^f(n)$ is now an uncorrelated signal, a linear predictor used in this manner is often called a whitening filter.

In this chapter, the reflection coefficients k_p are computed for a K sample block of data, and they thus are fixed over this block. In Chapter 7, adaptivity will be first introduced to the k_p via the gradient adaptive lattice algorithm, such that a time-varying reflection coefficient, $k_p(n)$, may be adaptive updated on a per sample basis. Then in Chapter 10, the very powerful least squares adaptive lattice filter will be derived and investigated for adaptively updating the forward and backward reflection coefficients.

PROBLEMS

1. If the z-transform of $p(n)$ is given by $Z\{p(n)\} = P(z)$; then find $Z\{p(n + 3)\}$.

2. Given the two difference equations below

$$x_1(n) = x_2(n - 1); \qquad x_2(n) = a_1 x_1(n - 1) + a_2 x_2(n - 1) + G\delta(n).$$

   Find the closed form solution for $x_1(n)$, given that $x_1(n) = x_2(n) = 0$ for $n < 0$.

3. Consider the case in which a two-stage lattice filter is used for linear prediction. Find expressions for the *transversal* linear prediction filter coefficients, w_1 and w_2, in terms of the reflection coefficients k_1 and k_2 for the lattice.

4. Show that the following relation used in deriving (3.2.17) is true:

$$k_p \varepsilon_{p-1} = \phi_x(p) - \mathbf{p}_{p-1}^T \bar{\mathbf{w}}_{p-1},$$

   where all quantities are as defined in Section 3.2.

5. Using Durbin's algorithm, compute the optimal $N = 2$ length linear prediction filter, as well as the reflection coefficients for the data having the following autocorrelation lags:

$$\phi_x(0) = 10, \qquad \phi_x(1) = 8, \qquad \phi_x(2) = 7.$$

6. Rework P2.4 using Durbin's algorithm, showing that this method and the direct matrix inverse give the same answer.

7. Rework P2.6 for the Gauss–Markov data signal model using Durbin's algorithm to compute the optimal $N = 2$ coefficient linear prediction filter.

8. Consider a data signal having an autoccorrelation function given by

$$\phi_x(m) = \sigma_x^2 \cos\left(\frac{m\pi}{4}\right) + \sigma_v^2 \delta(m).$$

(a) What are properties of the physical signals corresponding to this autocorrelation function?

(b) Using Durbin's algorithm, find the optimal $N = 1$ and 2 coefficient filters.

(c) What happens to these solutions as $\sigma_v^2 \to 0$ but σ_x^2 remains nonzero?

(d) What happens to the solutions as $\sigma_x^2 \to 0$, but σ_v^2 remains nonzero?

REFERENCES

1. N. Levinson, "The Wiener RMS (Root Mean Square) Error Criteria in Filter Design and Prediction," *J. Math. Phys.*, vol. 25, pp. 261–278, 1947.

2. J. Durbin, "Efficient Estimation of Parameters in Moving Average Models," *Biometrika*, vol. 46, pp. 306–316, 1959.

3. J.G. Proakis, *Digital Communications*, McGraw-Hill, New York, 1983.

4. B.S. Atal and S.L. Hanauer, "Speech Analysis and Synthesis by Linear Prediction," *J. Acous. Soc. Am.*, vol. 50, no. 2, pp. 637–644, August 1971.

5. J.D. Markel and A.H. Gray, *Linear Prediction of Speech*, Springer-Verlag, New York, 1975.

6. J.D. Markel and A.H. Gray, "On Autocorrelation Equations as Applied to Speech Analysis," *IEEE Trans. Audio and Electroacoustics*, vol. AU-21, pp. 69–79, 1973.

7. P.L. Chu and D.G. Messerschmitt, "A Frequency Weighted Itakura-Saito Spectral Distance Measure," *IEEE Trans. on Acous., Speech, and Signal Processing*, vol. ASSP-30, pp. 545–560, August 1982.

8. L.R. Rabiner and R.W. Schafer, *Digital Processing of Speech Signals*, Prentice-Hall, Englewood Cliffs, NJ, 1978.

9. A.V. Oppenheim and R.W. Schafer, *Digital Signal Processing*, Prentice-Hall, Englewood Cliffs, NJ, 1975.

10. J. Makhoul, "Linear Prediction: A Tutorial," *Proceedings of the IEEE*, vol. 63, pp. 561–580, April 1975.

11. S.M. Kay and S.L. Marple, "Spectrum Analysis—A Modern Perspective," *Proceeding of the IEEE*, vol. 69, pp. 1380–1419, November 1981.

12. J. Makhoul, "Stable and Efficient Lattice Methods for Linear Prediction," *IEEE Trans. on Acous., Speech, and Signal Processing*, vol. ASSP-25, pp. 423–428, October 1977.

13. S.T. Alexander, "A Simple Noniterative Speech Excitation Algorithm Using LPC Residual," *IEEE Trans. on Acous., Speech, and Signal Processing*, vol. ASSP-33, pp. 432–434, April 1985.

14. D.T. Paris and F.K. Hurd, *Basic Electromagnetic Theory*, McGraw-Hill, New York, 1969.

15. S. Haykin, ed., *Array Signal Processing*, Prentice-Hall, Englewood Cliffs, NJ, 1985.

CHAPTER 4
The Method of Steepest Descent

4.1 Introduction

In Chapter 2, a basic matrix inverse property was used to solve for the optimal mean square error (MSE) weights $\mathbf{w}_N^*$, producing the solution

$$\mathbf{w}_N^* = \mathbf{R}_{NN}^{-1}\mathbf{p}_N. \tag{2.3.6}$$

However, premultiplication by the inverse is not the only method for solving the simultaneous equations that led to $\mathbf{w}_N^*$ in (2.3.6). Actually, the procedure in (2.3.6) is of more theoretical than practical interest. Indeed, the subject of Chapter 3 was Durbin's algorithm, which is often used in practice to solve for the $\mathbf{w}_N$. However, there is another algorithm, known as the method of steepest descent (or simply steepest descent), which provides an iterative procedure for obtaining the same set of $\mathbf{w}_N^*$ as in (2.3.6). Since this iterative procedure is the basis for many currently used adaptive signal-processing techniques, the implications of the method of steepest descent will be explored in detail in this chapter. Steepest descent will be seen to lead directly to the popular least mean squares (LMS) adaptive algorithm, which has been widely implemented in actual systems. While the method of steepest descent itself is rarely used in actual adaptive signal processing systems, approximations to it are quite frequently implemented. Therefore, mathematical methods for analyzing steepest descent form the basis for investigating the theoretical properties of many adaptive techniques. The implications of Durbin's algorithm upon adaptive filter structures is deferred until Chapter 7 on the gradient adaptive lattice.

The method of steepest descent is often discussed in texts addressing numerical solution of simultaneous equations. The text by Wilde [1] is introductory and has many examples related to physical situations. The work by Hestenes [2] has a more mathematical orientation and is somewhat more advanced. The geometrical interpretations of the method of steepest descent is developed very well in the text by Pierre [3]. Steepest descent, as well as other computational methods for solving linear algebraic equations, is also developed in the computationally oriented text by Faddeev and Faddeeva [4].

Perhaps the greatest utility of the method of steepest descent is in the prediction of stochastic time series. A classical work in this area is the original paper by Robbins and Munro [5]; which addresses stochastic approximation. Many of the gain-normalized adaptive signal processing algorithms may be considered as variants of the Robbins–Munro method. Another tutorial work concerning gradient-based methods is that by Sakrison [6].

Much of the analytical work done in gradient based adaptive filtering (such as the LMS algorithm) also contains analyses of the method of steepest descent for comparison purposes. The text by Widrow and Stearns [20] covers many aspects of gradient-based methods at the introductory level. Also at the introductory level are three excellent works in this area by Widrow [7], Widrow and McCool [8], and Treichler [9], and at the somewhat more advanced level, the paper by Senne and Horowitz [10] is very explanatory.

Finally, use of the gradient algorithm is not limited to transversal filter applications. Cadzow [11] displays how gradient-based methods may be used to solve the nonlinear equations resulting from problems in recursive filter design. Additionally, Bard [12] has compiled performance evaluations of various gradient-based methods for solving other sets of nonlinear equations. A very complete examination of the method of steepest descent and the resulting gradient-based methods, as well as more sophisticated techniques of least squares estimation, are contained in the text by Eykhoff [13].

4.2 Iterative Solution of the Normal Equations

The mathematical development of the method of steepest descent is easily seen from the viewpoint of a geometrical approach using the MSE surface. In this text, points on the three-dimensional surface will usually be denoted with the first two coordinate values written in vector form, since, indeed, they are the coefficients of the filter vector that produces the particular MSE given by the third coordinate. As seen in Chapter 2, an MSE surface was associated with the $\mathbf{w}_N$-plane; that is, for every selection of a filter $\mathbf{w}_N$, there corresponds only one point on the MSE surface. Suppose the situation is like that in Figure 4.1, in which an initial filter setting, $\mathbf{w}_N(0)$, is arbitrarily chosen in the $\mathbf{w}_N$-plane. This is illustrated for the case of $N = 2$, but the concept extends to any general N. At the corresponding point on the MSE surface, $[\mathbf{w}_N(0), \varepsilon(0)]$, there is a specific orientation to the surface that may be described by the directional

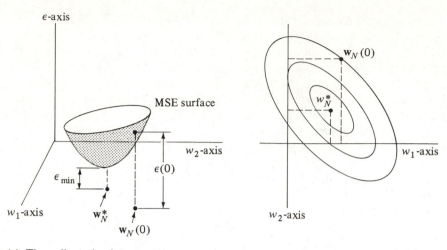

(a) Three dimension interpretation   (b) MSE contours in $\mathbf{w}_N$-plane

Figure 4.1   Initial choice $\mathbf{w}_N(0)$ and relation to MSE contours.

derivatives of the surface at that point. These directional derivatives quantify the rate of change of the MSE surface with respect to the $\mathbf{w}_N$ coordinate axes. That is, at the point on the surface $[\mathbf{w}_N(0), \varepsilon(0)]$, there is an instantaneous "slope" to the surface along a line parallel to the w_1-axis and an instantaneous slope along a line parallel to the w_2-axis. These instantaneous slopes of the surface have values defined by the directional derivatives, $\partial\varepsilon/\partial w_1$ and $\partial\varepsilon/\partial w_2$, which are then evaluated at the point $\varepsilon = \varepsilon(0)$. For simplicity, $\varepsilon = \varepsilon(n)$ denotes an evaluation of the "height" of the MSE surface ε at the filter vector location $\mathbf{w}_N(n)$.

The *gradient* of the error surface is defined as the vector of these directional derivatives:

$$\nabla_{\mathbf{w}}[\varepsilon] = \frac{\partial\varepsilon}{\partial\mathbf{w}_N}. \tag{4.2.1}$$

Thus, evaluation of the gradient at $\varepsilon = \varepsilon(0)$ for the case $N = 2$ becomes

$$\left[\frac{\partial\varepsilon}{\partial\mathbf{w}_N}\right]_{\varepsilon=\varepsilon(0)} = \begin{bmatrix} \dfrac{\partial\varepsilon}{\partial w_1} \\[2mm] \dfrac{\partial\varepsilon}{\partial w_2} \end{bmatrix}_{\varepsilon=\varepsilon(0)}. \tag{4.2.2}$$

In (4.2.2), the result is shown for $N = 2$, but the extension to arbitrary N is straightforward. By definition, the gradient in (4.2.1) points in the direction of the maximum rate of change of the surface at the point on the surface $[\mathbf{w}_N(0), \varepsilon(0)]$. The projection of this gradient onto the MSE contours is shown in Figure 4.2.

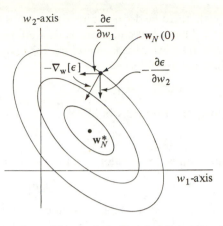

Figure 4.2   Construction of first update direction from directional derivatives.

Now suppose, in Figure 4.1., a "mental marble" were placed at this point on the surface, which is directly above $\mathbf{w}_N(0)$. If the marble were released, it would roll down toward the minimum of the surface, and it would initially roll in a direction opposite to the direction of maximum rate of change of the surface. This direction is opposite that of the gradient or, equivalently, this direction is that of "steepest descent." Using the geometrical results from Chapter 2 that the MSE surface is "bowl shaped," then the marble may be interpreted as rolling toward the bottom of the bowl. The marble might over-shoot or roll back and forth for a while, but provided the shape of the bowl does not change, the marble will eventually settle at the "bottom of the bowl."

This is a pleasing geometrical (and perhaps gravitational!) concept, but it is completely analogous to the mechanics of the method of steepest descent. The concepts of MSE surface and gradient of the MSE surface are necessary for an intuitive understanding of how the iterative steepest descent solution converges to $\mathbf{w}_N^*$. The method of steepest descent solves the normal equations via a mathematical method similar in concept to the qualitative approach of rolling the marble toward the bottom of the bowl. If "snapshots" at discrete time intervals were taken of the position of the marble, then the position of the marble would move in a stepwise fashion toward the minimum of the MSE surface. From each new discrete time position, the marble would subsequently roll toward the global minimum in a direction dependent upon the MSE surface gradient at that position. The mathematical equivalent of this property is that given a position on the MSE surface at "time" n (more accurately, at iteration n of the algorithm), then the marble position at time $n + 1$ is achieved by moving along a direction opposite to the gradient. This may be formally written as follows. If the initial position on the MSE surface were $[\mathbf{w}_N(0), \varepsilon(0)]$, then the next position $[\mathbf{w}_N(1), \varepsilon(1)]$ would be related to the initial position by

$$\begin{bmatrix} \mathbf{w}_N(1) \\ \varepsilon(1) \end{bmatrix} = \begin{bmatrix} \mathbf{w}_N(0) - \mu \nabla_{\mathbf{w}}[\varepsilon(0)] \\ \varepsilon(0) - \delta \end{bmatrix}.$$

where μ is some constant (to be defined directly) and δ is the change in MSE. Since the solution to the normal equations is desired, the upper partition in the above equation will be examined. Therefore, the position in the $\mathbf{w}_N$-plane of the marble at time 1 is given by

$$\mathbf{w}_N(1) = \mathbf{w}_N(0) - \mu\nabla_{\mathbf{w}}[\varepsilon(0)].$$

Similarly, the $\mathbf{w}_N$ position at time 2 would be

$$\mathbf{w}_N(2) = \mathbf{w}_N(1) - \mu\nabla_{\mathbf{w}}[\varepsilon(1)],$$

and the general recursion is immediately seen to be given by

$$\mathbf{w}_N(n + 1) = \mathbf{w}_N(n) - \mu\nabla_{\mathbf{w}}[\varepsilon(n)]. \tag{4.2.3}$$

Therefore, (4.2.3) is the mathematical algorithm corresponding to rolling the marble toward the bottom of the bowl. The $\mathbf{w}_N(n)$ weight vector in (4.2.3) is the set of filter coefficient values at the nth iteration of the algorithm. This recursion expresses the new weight vector as a function of the old weight vector, plus a correction term dependent upon the properties of the error surface at the old weight vector position.

The present purpose of the method of steepest descent is to solve the normal equations for $\mathbf{w}_N^*$, which is simply the solution to a set of linear equations. Many optimization problems not associated with signal processing can thus be solved in this manner [1–3]. The application to minimum MSE filtering occurs by finding the gradient of the MSE surface as a function of the $\mathbf{R}_{NN}$ matrix and the $\mathbf{p}_N$ vector. This procedure thus relates the solution of the matrix equation (2.3.5) to the signal-processing environment. To find this expression, recall (2.2.17), which gives the MSE as a function of the predictor weights $\mathbf{w}_N$:

$$\varepsilon(\mathbf{w}_N) = \sigma_d^2 - 2\mathbf{w}_N^T\mathbf{p}_N + \mathbf{w}_N^T\mathbf{R}_{NN}\mathbf{w}_N. \tag{4.2.4}$$

In the method of steepest descent, the weight vector will vary as a function of the iteration number n, and therefore will be denoted by $\mathbf{w}_N(n)$. Thus, the MSE is also a function of the iteration n and (4.2.4) becomes:

$$\varepsilon(n) = \sigma_d^2 - 2\mathbf{w}_N^T(n)\mathbf{p}_N + \mathbf{w}_N^T(n)\mathbf{R}_{NN}\mathbf{w}_N(n), \tag{4.2.5}$$

where the shortened notation $\varepsilon(n) = \varepsilon[\mathbf{w}_N(n)]$ has been used. Using (4.1.1), the gradient of $\varepsilon(n)$ is now given by

$$\nabla_{\mathbf{w}}[\varepsilon(n)] = -2\mathbf{p}_N + 2\mathbf{R}_{NN}\mathbf{w}_N(n). \tag{4.2.6}$$

Note that (4.2.5) and (4.2.6) are valid for any value of n.

Substituting (4.2.6) for the gradient into (4.2.3) produces the following form for the steepest descent recursion:

$$\mathbf{w}_N(n + 1) = \mathbf{w}_N(n) + 2\mu[\mathbf{p}_N - \mathbf{R}_{NN}\mathbf{w}_N(n)]. \tag{4.2.7}$$

Since μ is a constant, it is common to define

$$\alpha = 2\mu, \tag{4.2.8}$$

which gives the desired form for the method of steepest descent recursion

$$\mathbf{w}_N(n + 1) = \mathbf{w}_N(n) + \alpha[\mathbf{p}_N - \mathbf{R}_{NN}\mathbf{w}_N(n)]. \qquad (4.2.9)$$

Note that (4.2.9) may also be expressed as

$$\mathbf{w}_N(n + 1) = [\mathbf{I}_{NN} - \alpha\mathbf{R}_{NN}]\mathbf{w}_N(n) + \alpha\mathbf{p}_N, \qquad (4.2.10)$$

where $\mathbf{I}_{NN}$ is the $N \times N$ identity matrix.

Therefore, the method of steepest descent could be applied to the minimum MSE filtering problem as follows. Estimates of the autocorrelation matrix $\mathbf{R}_{NN}$ and the cross-correlation vector $\mathbf{p}_N$ must first be computed from the available data. This could be done by using estimators such as (2.2.11) and (2.2.15) for the required correlations. These estimates, $\hat{\mathbf{R}}_{NN}$ and $\hat{\mathbf{p}}_N$, could then be used in (4.2.10) and successive iterations performed. Once $\hat{\mathbf{R}}_{NN}$ and $\hat{\mathbf{p}}_N$ have been estimated, the problem is once more solution of a matrix equation,

$$\hat{\mathbf{R}}_{NN}\mathbf{w}_N = \hat{\mathbf{p}}_N,$$

and steepest descent may be used. Although this procedure is rarely done in practice (due to computational complexity), there is nothing from a mathematical standpoint to preclude its use. In fact, it will be seen in Chapter 5 that many of the properties of the LMS algorithm can be related to the properties of the steepest descent recursion (4.2.10).

Convergence to $\mathbf{w}_N^*$

Equation (4.2.10) is the basic recursion defining the method of steepest descent. However, it is necessary to prove that the $\mathbf{w}_N(n)$ solution obtained using this iterative method is the same minimum MSE vector $\mathbf{w}_N^*$ obtained by the matrix inverse method of (2.3.6). To prove this, it is necessary to show that the $\mathbf{w}_N(n)$ sequence obtained by successive iterations of (4.2.10) closely approach, or *converge* to the optimum $\mathbf{w}_N^*$. Mathematically, it is necessary to prove

$$\lim_{n \to \infty} \mathbf{w}_N(n) = \mathbf{w}_N^*, \qquad (4.2.11)$$

or, equivalently,

$$\lim_{n \to \infty} \mathbf{v}_N(n) = \mathbf{0}_N, \qquad (4.2.12)$$

where from (2.5.3)

$$\mathbf{v}_N(n) = \mathbf{w}_N(n) - \mathbf{w}_N^* \qquad (4.2.13)$$

is the difference vector between the weights found by steepest descent and the optimal weights. As was displayed in Chapter 2, approaching the problem using the concept of difference vector often leads to a more tractable mathematical development, but still preserves the information about the convergence of $\mathbf{w}_N(n)$ to $\mathbf{w}_N^*$. Note also that the limit in (4.2.11) implies that each of the weight components, $w_i(n)$, converge to the corresponding minimum MSE weight component; that is,

$$\lim_{n \to \infty} w_i(n) = w_i^* \tag{4.2.14}$$

for $1 \le i \le N$. This suggests immediately that

$$\lim_{n \to \infty} v_i(n) = 0; \qquad 1 \le i \le N. \tag{4.2.15}$$

As in Chapter 2, a coordinate rotation may also be employed to obtain a set of rotated, or *uncoupled*, difference weights, $\mathbf{v}_N'(n)$. The term uncoupled will be seen to accurately describe the solution process when the convergence properties of steepest descent are investigated. From (2.5.8) and (2.5.10), the $\mathbf{v}_N'(n)$ vector is given by premultiplying $\mathbf{v}_N(n)$ by the transpose of the matrix $\mathbf{M}_{NN}$ whose columns are the eigenvectors of $\mathbf{R}_{NN}$:

$$\mathbf{v}_N'(n) = \mathbf{M}_{NN}^T \mathbf{v}_N(n). \tag{4.2.16}$$

It is easy to show that the elements of $\mathbf{v}_N'(n)$ are made up of linear combinations of the $v_i(n)$. Since, in the limit, each of the $v_i(n)$ vanishes, this implies that each of the $v_i'(n)$ vanishes, as well, or

$$\lim_{n \to \infty} \mathbf{v}_N'(n) = \mathbf{0}_N. \tag{4.2.17}$$

Therefore, proof of convergence of the steepest descent weights to the optimal $\mathbf{w}_N^*$ is equivalent to proving convergence of $\mathbf{v}_N'(n)$ to zero. Since convergence proofs are critical to the analysis of adaptive signal processing algorithms, this topic will be explored in detail in the next section. Whereas the specific analysis considered is for the iterative solution to a matrix equation, it will be seen in later work that this analytical approach is the basis of convergence analysis for the popular gradient-based LMS algorithm.

4.3 Weight Vector Solutions

In this section, closed form solutions for the weight vector obtained using the method of steepest descent will be derived. In so doing, the linear transformations of Chapter 2 will be employed to simplify the analysis.

To illustrate the utility of using the linear transformations, a sequence of examples will be examined, which will also provide an introduction to some basic analytical techniques. There are several ways of solving (4.2.10) that will provide closed form expressions for $\mathbf{w}_N(n)$. One method is to employ z-transform theory for solving difference equations. However, there is a simpler method of direct evaluation of the recursion (4.2.10) that offers potentially more insight into the physical process. In the current section, this direct evaluation will be examined, and the z-transform method will be investigated in a following section.

Direct evaluation of the difference equation

Recall the weight recursion (4.2.10) given by the method of steepest descent, repeated here:

$$\mathbf{w}_N(n + 1) = [\mathbf{I}_{NN} - \alpha\mathbf{R}_{NN}]\mathbf{w}_N(n) + \alpha\mathbf{p}_N. \tag{4.3.1}$$

Therefore, at $n = 0$,

$$\mathbf{w}_N(1) = [\mathbf{I}_{NN} - \alpha\mathbf{R}_{NN}]\mathbf{w}_N(0) + \alpha\mathbf{p}_N. \tag{4.3.2}$$

Similarly, for $n = 1$,

$$\mathbf{w}_N(2) = [\mathbf{I}_{NN} - \alpha\mathbf{R}_{NN}]\mathbf{w}_N(1) + \alpha\mathbf{p}_N. \tag{4.3.3}$$

If (4.3.2) for $\mathbf{w}_N(1)$ is then substituted into (4.3.3), a pattern is seen to be evolving, which may be verified by finding $\mathbf{w}_N(3)$. Continuing this process to the nth iteration would then produce

$$\mathbf{w}_N(n) = [\mathbf{I}_{NN} - \alpha\mathbf{R}_{NN}]^n\mathbf{w}_N(0) + \alpha\mathbf{p}_N \sum_{j=0}^{n-1} [\mathbf{I}_{NN} - \alpha\mathbf{R}_{NN}]^j. \tag{4.3.4}$$

Equation (4.3.4) is perfectly admissible for computing the weight vector at any iteration n. The values $\mathbf{p}_N$ and $\mathbf{R}_{NN}$ are known from (2.3.5) and may be substituted into (4.3.4) to compute the solution. However, this approach is not very viable, since matrix exponentials, plus a running sum of matrix exponentials, would have to be computed. In general these would be very laborious computational tasks. Moreover, (4.3.4) provides little insight into the convergence process of $\mathbf{w}_N(n)$ to $\mathbf{w}_N^*$. Indeed, it is difficult to determine from (4.3.4) whether the algorithm converges to any value or diverges completely. Therefore, a method is needed to simplify the interpretation of (4.3.4), and the linear transformations of Chapter 2 suggest one such approach.

In Chapter 2, the linear transformations of translation and rotation were applied to the $\mathbf{w}_N$-space to produce a very simple expression for the MSE as a function of the uncoupled weights $\mathbf{v}_N'(n)$. In this section, the same procedure will be applied to the method of steepest descent, and the resulting equations will very clearly display the requirements for convergence of the $\mathbf{w}_N(n)$ to the optimal $\mathbf{w}_N^*$. In order to illustrate some basic analytical techniques, this will be done in two steps. First a translation of the $\mathbf{w}_N$-axes will produce the $\mathbf{v}_N$-space, after which a rotation of the $\mathbf{v}_N$-axes will produce the uncoupled $\mathbf{v}_N'$ equations in the $\mathbf{v}_N'$-space.

The translation to the $\mathbf{v}_N$-space may be achieved very simply. Subtract $\mathbf{w}_N^*$ from both sides of (4.2.9) and use the definition of the difference vector from (4.2.13) to give

$$\mathbf{v}_N(n + 1) = \mathbf{v}_N(n) + \alpha[\mathbf{p}_N - \mathbf{R}_{NN}\mathbf{w}_N(n)]. \tag{4.3.5}$$

Equation (4.3.5) can be made a function of only the difference vector by adding and subtracting $\alpha \mathbf{R}_{NN} \mathbf{w}_N^*$ to the right-hand side, and then grouping terms:

$$\mathbf{v}_N(n + 1) = \mathbf{v}_N(n) - \alpha \mathbf{R}_{NN}[\mathbf{w}_N(n) - \mathbf{w}_N^*] + \alpha \mathbf{p}_N - \alpha \mathbf{R}_{NN} \mathbf{w}_N^*. \qquad (4.3.6)$$

But from (4.2.13), the quantity in brackets in simply $\mathbf{v}_N(n)$; and from (2.3.5), $\mathbf{p}_N = \mathbf{R}_{NN} \mathbf{w}_N^*$, which gives the following desired form for the difference vector recursion:

$$\mathbf{v}_N(n + 1) = [\mathbf{I}_{NN} - \alpha \mathbf{R}_{NN}] \mathbf{v}_N(n). \qquad (4.3.7)$$

The form (4.3.7) is much easier to evaluate than the direct solution (4.3.4). Expand (4.3.7) for $n = 0$:

$$\mathbf{v}_N(1) = [\mathbf{I}_{NN} - \alpha \mathbf{R}_{NN}] \mathbf{v}_N(0),$$

and for $n = 1$:

$$\mathbf{v}_N(2) = [\mathbf{I}_{NN} - \alpha \mathbf{R}_{NN}] \mathbf{v}_N(1) = [\mathbf{I}_{NN} - \alpha \mathbf{R}_{NN}]^2 \mathbf{v}_N(0).$$

It is therefore easy to see that the general solution will be given by

$$\mathbf{v}_N(n) = [\mathbf{I}_{NN} - \alpha \mathbf{R}_{NN}]^n \mathbf{v}_N(0). \qquad (4.3.8)$$

Equation (4.2.13) therefore gives

$$\mathbf{w}_N(n) = \mathbf{w}_N^* + [\mathbf{I}_{NN} - \alpha \mathbf{R}_{NN}]^n \mathbf{v}_N(0). \qquad (4.3.9)$$

While (4.3.9) is an improvement over (4.3.4) for determining the dynamic behavior of the $\mathbf{w}_N(n)$, there still is one difficulty. Even though the inclusion of $\mathbf{w}_N^*$ in (4.3.9) is evident, it is still not clear whether the second term on the right-hand side vanishes for $n \to \infty$. Insight into the computational difficulty may be gained by taking a simple example for $N = 2$ and expanding (4.3.9) for each component at the $n = 1$st iteration:

$$\begin{bmatrix} w_1(1) \\ w_2(1) \end{bmatrix} = \begin{bmatrix} w_1^* \\ w_2^* \end{bmatrix} + \left\{ \begin{bmatrix} 1 & 0 \\ 0 & 1 \end{bmatrix} - \alpha \begin{bmatrix} \phi_x(0) & \phi_x(1) \\ \phi_x(1) & \phi_x(0) \end{bmatrix} \right\} \begin{bmatrix} v_1(0) \\ v_2(0) \end{bmatrix}. \qquad (4.3.10)$$

Expanding (4.3.10) then produces the following equations for the weights:

$$w_1(1) = w_1^* + [1 - \alpha \phi_x(0)] v_1(0) - \alpha \phi_x(1) v_2(0)$$
$$w_2(1) = w_2^* + [1 - \alpha \phi_x(0)] v_2(0) - \alpha \phi_x(1) v_1(0). \qquad (4.3.11)$$

Note that each weight value affects the solution for the other weight because of the cross-coupling of $v_1(0)$ and $v_2(0)$. This situation is compounded if the general N-length vector $\mathbf{w}_N$ is examined, in which case the weight $w_1(n)$ is a function of all other weights $w_2(n), w_3(n), \ldots, w_N(n)$, the weight $w_2(n)$ is a function of $w_1(n), w_3(n), \ldots, w_N(n)$, and so forth.

This cross-coupling effectively prohibits a great deal of analytical insight into the dynamics of the $\mathbf{w}_N(n)$ from (4.3.9). Again, $\mathbf{w}_N(n)$ can indeed be calculated, but it is still not evident how the $\mathbf{w}_N(n)$ converge or if they converge at all. This analytical difficulty is a direct result of the fact that $[\mathbf{I}_{NN} - \alpha \mathbf{R}_{NN}]$

is not a diagonal matrix. However, in Chapter 2, it was seen that premultiplication of the MSE equation by the orthogonal modal matrix of $\mathbf{R}_{NN}$ aligned
the principal axes of the MSE contours with the coordinate axes in $\mathbf{v}'_N$-space.
Additionally, it was seen that the mathematical operation that produced this
alignment was an orthogonal transformation. Using this orthogonal transformation on the steepest descent recursion (4.3.1) will provide a similar
beneficial effect, transforming the *coupled* difference equations (4.3.10) into a
set of *uncoupled* difference equations. In the uncoupled difference equations,
each equation will be a function of only a single scalar weight, and each of
these scalar equations may therefore be solved in a very simple manner.

It should be mentioned that the translation of axes and the rotation of the
axes are linear operations, and the order of their application is immaterial.
Sometimes it is necessary to formulate analysis in the rotated-and-translated
space $\mathbf{v}'_N$, whereas at other times, it may not be necessary to make the complete
transition to this space. Of the two operations, the rotation of the axes (i.e.,
the orthogonal transformation) is the more important, since it uncouples the
vector difference equation and allows the solution of a set of scalar difference
equations. As an example of this approach, the convergence properties of the
solution given by the method of steepest descent will be examined next by
working in the uncoupled (rotated) $\mathbf{w}'_N$-space. This space is obtained by performing a rotation of the original $\mathbf{w}_N$-axes, but not translating the origin to
the optimal filter location. In the problems at the end of the chapter, the reader
is asked to develop the same final solution to this problem from the standpoint
of the rotated-and-translated $\mathbf{v}'_N$-space and demonstrate the equivalence of the
approaches.

Orthogonal transformations

In this section, an orthogonal transformation will be used to produce a set of
uncoupled difference equation that may then be solved in a simple manner.
A solution will be derived for $\mathbf{w}'_N(n)$, which is a function of the iteration number
n. If required by the application, this solution could then be rotated back to
the $\mathbf{w}_N(n)$-space to determine the actual weight solution. However, many times
the desired properties are evident from an examination of the $\mathbf{w}'_N(n)$ solution,
such that recovery of the $\mathbf{w}_N(n)$ is not necessary.

To derive the solution for the uncoupled weights, substitute the similarity
transformation (2.5.10) for the autocorrelation matrix into the weight recursion (4.2.10). This gives

$$\mathbf{w}_N(n+1) = \mathbf{M}_{NN}[\mathbf{I}_{NN} - \alpha\mathbf{\Lambda}_{NN}]\mathbf{M}_{NN}^T\mathbf{w}_N(n) + \alpha\mathbf{p}_N, \qquad (4.3.12)$$

where the property $\mathbf{I}_{NN} = \mathbf{M}_{NN}\mathbf{M}_{NN}^T$ for the orthogonal matrix has been used.
The uncoupled weights $\mathbf{w}'_N(n)$ are next defined by the transformation

$$\mathbf{w}'_N(n) = \mathbf{M}_{NN}^T\mathbf{w}_N(n). \qquad (4.3.13)$$

Note that the transformation in (4.3.13) also defines a set of optimal uncoupled weights:

$$\mathbf{w}_N'^* = \mathbf{M}_{NN}^T \mathbf{w}_N^*. \tag{4.3.14}$$

To obtain a solution for the uncoupled weights, premultiply (4.3.12) by $\mathbf{M}_{NN}^T$ and substitute (4.3.13):

$$\mathbf{w}_N'(n+1) = \mathbf{M}_{NN}^T \mathbf{M}_{NN}[\mathbf{I}_{NN} - \alpha\Lambda_{NN}]\mathbf{w}_N'(n) + \alpha\mathbf{M}_{NN}^T \mathbf{p}_N$$
$$= [\mathbf{I}_{NN} - \alpha\Lambda_{NN}]\mathbf{w}_N'(n) + \alpha\mathbf{p}_N', \tag{4.3.15}$$

where the following definition in (4.3.15) has been used

$$\mathbf{p}_N' = \mathbf{M}_{NN}^T \mathbf{p}_N. \tag{4.3.16}$$

Therefore from (4.3.14)

$$\mathbf{w}_N'^* = \mathbf{M}_{NN}^T \mathbf{w}_N^* = \mathbf{M}_{NN}^T \mathbf{R}_{NN}^{-1} \mathbf{p}_N. \tag{4.3.17}$$

However, the special properties of the orthogonal modal matrix allow the inverse of $\mathbf{R}_{NN}$ to be written as [14–16]

$$\mathbf{R}_{NN}^{-1} = (\mathbf{M}_{NN}\Lambda_{NN}\mathbf{M}_{NN}^T)^{-1} = \mathbf{M}_{NN}\Lambda_{NN}^{-1}\mathbf{M}_{NN}^T. \tag{4.3.18}$$

Therefore, substituting (4.3.18) into (4.3.17) gives

$$\mathbf{w}_N'^* = \mathbf{M}_{NN}^T \mathbf{M}_{NN}\Lambda_{NN}^{-1}\mathbf{M}_{NN}^T \mathbf{p}_N = \Lambda_{NN}^{-1}\mathbf{p}_N'. \tag{4.3.19}$$

Equation (4.3.19) is very similar to (2.3.6), except that it is in terms of the uncoupled weight vector. Since Λ_{NN} is a diagonal matrix, its inverse is also diagonal, with elements

$$(\Lambda_{NN}^{-1})_{ii} = \frac{1}{\lambda_i}. \tag{4.3.20}$$

Now an expression can be found for each component of $\mathbf{w}_N'^*$ by expanding (4.3.19), producing

$$\mathbf{w}_N'^* = \begin{bmatrix} \dfrac{1}{\lambda_1} & 0 & \cdots & 0 \\ 0 & \dfrac{1}{\lambda_2} & & 0 \\ \cdots & \cdots & & \cdots \\ 0 & 0 & \cdots & \dfrac{1}{\lambda_N} \end{bmatrix} \begin{bmatrix} p_1' \\ p_2' \\ \cdots \\ p_N' \end{bmatrix}, \tag{4.3.21}$$

from which the individual $w_i'^*$ are seen to be given by

$$w_i'^* = \frac{p_i'}{\lambda_i}. \tag{4.3.22}$$

Since $[\mathbf{I}_{NN} - \alpha\Lambda_{NN}]$ is diagonal, the recursion (4.3.16) may be expanded as

$$\begin{bmatrix} w_1'(n+1) \\ w_2'(n+1) \\ \cdots \\ w_N'(n+1) \end{bmatrix} = \begin{bmatrix} 1-\alpha\lambda_1 & 0 & \cdots & 0 \\ 0 & 1-\alpha\lambda_2 & \cdots & 0 \\ \cdots & \cdots & & \cdots \\ 0 & 0 & \cdots & 1-\alpha\lambda_N \end{bmatrix} \begin{bmatrix} w_1'(n) \\ w_2'(n) \\ \cdots \\ w_N'(n) \end{bmatrix}$$

$$+ \begin{bmatrix} \alpha p_1' \\ \alpha p_2' \\ \cdots \\ \alpha p_N' \end{bmatrix}. \tag{4.3.23}$$

Note that each $w_i'(n+1)$ is only for a function $w_i'(n)$ and is not a function of any of the other $w_j'(n)$ for $j \neq i$. This is exactly the property needed to write the system of equations in (4.3.23) as a set of N scalar-uncoupled equations. These uncoupled equations are given by

$$w_i'(n+1) = (1-\alpha\lambda_i)w_i'(n) + \alpha p_i'; \qquad 1 \leq i \leq N. \tag{4.3.24}$$

Since the same scalar difference equation holds for all $w_i'(n)$, then (4.3.24) may be solved for the general case and the entire set of solutions obtained for $1 \leq i \leq N$.

Since (4.3.24) is a relatively simple form, it will be intructive to solve it in two ways: (1) by direct evaluation of the difference equation, and (2) by using z-transforms. Both methods should produce the same answer, of course, which will provide a verification of the solution techniques. Moreover, this approach will provide additional experience in basic analytical techniques.

Weight solution via direct evaluation

The recursion (4.3.24) may be evaluated for successive values of n, and a general pattern may be recognized. Expand (4.3.24) for $n = 0$,

$$w_i'(1) = (1-\alpha\lambda_i)w_i'(0) + \alpha p_i',$$

and for $n = 1$,

$$w_i'(2) = (1-\alpha\lambda_i)w_i'(1) + \alpha p_i',$$

$$= (1-\alpha\lambda_i)[(1-\alpha\lambda_i)w_i'(0) + \alpha p_i'] + \alpha p_i'$$

$$= (1-\alpha\lambda_i)^2 w_i'(0) + \alpha p_i' \sum_{j=0}^{1}(1-\alpha\lambda_i)^j.$$

The pattern emerging can be seen from this last result, and successive substitutions would show this pattern to be

$$w_i'(n) = (1-\alpha\lambda_i)^n w_i'(0) + \alpha p_i'\left[\sum_{j=0}^{n-1}(1-\alpha\lambda_i)^j\right]. \tag{4.3.25}$$

Since the factor $(1-\alpha\lambda_i)$ appears so frequently in work to follow, use the substitution

$$\gamma_i = 1 - \alpha\lambda_i. \tag{4.3.26}$$

Substituting (4.3.26) into (4.3.25) then gives the following solution for the $w_i'(n)$:

$$w_i'(n) = \gamma_i^n w_i'(0) + \alpha p_i' \sum_{j=0}^{n-1} \gamma_i^j. \tag{4.3.27}$$

The summation in (4.3.27) is immediately seen to be the familiar geometric series, which has the closed form expression

$$\sum_{j=0}^{n-1} \gamma_i^j = \frac{1 - \gamma_i^n}{1 - \gamma_i}. \tag{4.3.28}$$

Substituting this result into (4.3.27) then produces the desired form for the uncoupled weight solution:

$$w_i'(n) = \gamma_i^n w_i'(0) + \alpha p_i' \left[\frac{1 - \gamma_i^n}{1 - \gamma_i} \right]. \tag{4.3.29}$$

Solution by z-transforms

An alternate method of solution for the $w_i'(n)$ is obtained by employing z-transforms [17–19]. The difference equation for the $w_i'(n)$ is given by (4.3.24), using the substitution of (4.3.26):

$$w_i'(n + 1) = \gamma_i w_i'(n) + \alpha p_i'. \tag{4.3.30}$$

Computing the z-transforms of each term in (4.3.30) gives the corresponding z-domain expression:

$$z W_i'(z) - z w_i'(0) = \gamma_i W_i'(z) + Z\{\alpha p_i'\}, \tag{4.3.31}$$

where $W_i'(z)$ is the z-transform of $w_i'(n)$,

$$Z\{w_i'(n)\} = W_i'(z) \tag{4.3.32}$$

and $w_i'(0)$ is the initial condition. The last term on the right-hand side of (4.3.31) has a z-transform given by

$$Z\{\alpha p_i'\} = \alpha p_i' \frac{1}{(1 - z^{-1})}. \tag{4.3.33}$$

Using (4.3.33) in (4.3.31) and then grouping terms produces

$$W_i'(z) = \frac{w_i'(0)}{(1 - \gamma_i z^{-1})} + \frac{\alpha p_i' z^{-1}}{(1 - z^{-1})(1 - \gamma_i z^{-1})}. \tag{4.3.34}$$

Equation (4.3.34) is the solution for $W_i'(z)$, and the corresponding solution for $w_i'(n)$ requires the inverse z-transform of each term in (4.3.34). This is easily done using standard z-transform techniques [17, 18] and gives the complete result for $w_i'(n)$:

$$w_i'(n) = \gamma_i^n w_i'(0) + \alpha p_i' \left[\frac{1 - \gamma_i^n}{1 - \gamma_i} \right], \tag{4.3.35}$$

which is seen to be exactly the same as (4.3.29).

Therefore, the same solution for the uncoupled weights may be derived via either direct evaluation or the z-transform method. An ability to use either method will be very valuable, since there will be instances in which one method will be more applicable than the other. The next section will examine the implications and properties of the solutions (4.3.29) and (4.3.35).

4.4 Convergence Properties of Steepest Descent

At this point, the questions concerning the convergence of the steepest descent solution (4.2.10) to the optimal $\mathbf{w}_N^*$ may be answered. However, it will be seen later that a great deal more information is available concerning the dynamics of the weight convergence process. To investigate the absolute convergence of steepest descent, consider the behavior of (4.3.29) as $n \to \infty$. It is seen that γ_i is raised to an increasing positive power as time increases. Therefore, one requirement for (4.3.29) to produce a bounded solution for $w_i'(n)$ is that the absolute value of γ_i is less than unity for all $1 \leq i \leq N$:

$$|\gamma_i| < 1. \tag{4.4.1}$$

The condition (4.4.1) imposes a bound on the gain parameter α, which from (4.3.26) is seen to be

$$|1 - \alpha\lambda_i| < 1, \tag{4.4.2}$$

which is equivalent to

$$-1 < (1 - \alpha\lambda_i) < 1. \tag{4.4.3}$$

It is easy to show that the bounds in (4.4.3) are equivalent to

$$0 < \alpha < \frac{2}{\lambda_i}. \tag{4.4.4}$$

It is required that (4.4.4) hold for any $1 \leq i \leq N$, and therefore the lowest upper bound that any λ_i enforces must be found. This occurs when $\lambda_i = \lambda_{\max}$, the maximum eigenvalue of the autocorrelation matrix $\mathbf{R}_{NN}$ and the range of α to insure convergence is given by

$$0 < \alpha < \frac{2}{\lambda_{\max}}. \tag{4.4.5}$$

For α within the range expressed in (4.4.5), then $0 < |\gamma_i| < 1$ and using (4.3.29) shows

$$\lim_{n \to \infty} w_i'(n) = \frac{p_i'}{\lambda_i}. \tag{4.4.6}$$

Note that this is exactly the result for $w_N'^*$, the uncoupled minimum MSE weight solution obtained from (4.3.22). Therefore, the weights obtained via the method of steepest descent do converge to the same set of minimum MSE weights that solve the normal equations, provided that α is within the range expressed in (4.4.5).

Dynamics of weight convergence process

Now that a proof of convergence has been obtained, an examination of the specific process by which the steepest descent solution converges to the optimal w_N^* will be valuable. Among the valid considerations at this point might be the time constants of the convergence process, or perhaps whether the convergence process is montonic or oscillatory. In this section, the convergence process will be examined using the weight solutions in the uncoupled v_N'-space, which will provide another viewpoint for the analytical process.

From (4.2.13) the expression for the weight difference vector is given by

$$v_N(n) = w_N(n) - w_N^*, \tag{4.2.13}$$

and from (4.2.16), the uncoupled weight difference vector is given by the transformation

$$v_N'(n) = M_{NN}^T v_N(n). \tag{4.2.16}$$

Substituting (4.2.16) into (4.2.13) and using (4.3.13) then produces

$$v_N'(n) = w_N'(n) - w_N'^*. \tag{4.4.7}$$

Substituting (4.3.22) and (4.3.29) into (4.4.7), the ith component $v_i'(n)$ is thus given by

$$v_i'(n) = \gamma_i^n w_i'(0) + \alpha p_i' \left[\frac{1 - \gamma_i^n}{1 - \gamma_i} \right] - \frac{p_i'}{\lambda_i}, \tag{4.4.8}$$

which after intermediate algebra produces

$$v_i'(n) = \left[w_i'(0) - \frac{p_i'}{\lambda_i} \right] (1 - \alpha \lambda_i)^n. \tag{4.4.9}$$

Note that the bracketed term in (4.4.9) is a constant once the initial weights $w_i'(0)$ are chosen, and that this constant is equal to $v_i'(0)$. Therefore, (4.4.9) may be written more simply as

$$v_i'(n) = (1 - \alpha \lambda_i)^n v_i'(0). \tag{4.4.10}$$

From an examination of (4.4.10), it is seen that for the case of $v_i'(0) \neq 0$, the uncoupled difference weights converge exponentially to zero, with a time constant dependent upon the size of α relative to the eigenvalues of R_{NN}. Recalling the definition of the $v_i'(n)$, this implies the actual weight vector $w_N(n)$ from steepest descent converges to the optimal w_N^* in the same manner. An important point to note from (4.4.10) is that the $v'_i(n)$ will approach zero regardless of the initial value of $v'_i(0)$, which is equivalent to stating that the $\mathbf{w}_N(n)$ will converge to $\mathbf{w}_N^*$ regardless of the initial value $\mathbf{w}_N(0)$. Therefore, regardless of how little knowledge is possessed about the correct optimal $\mathbf{w}_N^*$, (4.4.10) guarantees $\mathbf{w}_N(n)$ will converge to $\mathbf{w}_N^*$, provided α is within the range established by (4.4.5). This is indeed a very important property of the method of steepest descent and will be seen to be preserved in the gradient-based adaptive algorithms explored in later chapters.

Convergence examples

Some examples are valuable at this time to illustrate the preceding concepts. Suppose the systems identification problem (see Figure 1.1) is attempted and the second-order statistics of the input $x(n)$ and output $d(n)$ are computed and found to be

$$\hat{\phi}_x(0) = 1.0, \qquad \hat{\phi}_x(1) = 0.8$$
$$\hat{\phi}_{xd}(0) = 0.8, \qquad \hat{\phi}_{xd}(1) = 0.5. \tag{4.4.11}$$

It is desired to find the best $N = 2$ model of the unknown system, and this requires solving

$$\hat{\mathbf{R}}_{NN}\mathbf{w}_N(n) = \hat{\mathbf{p}}_N, \tag{4.4.12}$$

where

$$\hat{\mathbf{R}}_{NN} = \begin{bmatrix} 1.0 & 0.8 \\ 0.8 & 1.0 \end{bmatrix}, \qquad \hat{\mathbf{p}}_N = \begin{bmatrix} 0.8 \\ 0.5 \end{bmatrix}. \tag{4.4.13}$$

Although (4.4.12) is a simple 2×2 matrix equation and could be solved directly, using steepest descent will illustrate principles that will be very useful for more complicated problems.

First, it is simple to show that the eigenvalues of $\hat{\mathbf{R}}_{NN}$ are $(\lambda_1, \lambda_2) = (1.8, 0.2)$, and the normalized eigenvectors are

$$\mathbf{m}_1 = \frac{1}{\sqrt{2}}[1, 1]^T, \qquad \mathbf{m}_2 = \frac{1}{\sqrt{2}}[1, -1]^T.$$

Thus, for stability of steepest descent, $\alpha < 2/1.8$. For the current example, choose $\alpha = 1$ and let $\mathbf{w}_N(0) = 0$.

Equation (4.3.35) provides the simplest method for solving for the $w_i(n)$ trajectories. For the present example, (4.3.35) simplifies to

$$w'_i(n) = \frac{p'_i}{\lambda_i}[1 - (1 - \alpha\lambda_i)^n]. \tag{4.4.14}$$

Using (4.3.16) it is easy to show that

$$\mathbf{p}'_N = \frac{1}{\sqrt{2}}[1.3, 0.3]^T,$$

from which

$$w_1'(n) = \frac{1.3}{1.8\sqrt{2}}[1 - (-0.8)^n]$$

$$w_2'(n) = \frac{0.3}{0.2\sqrt{2}}[1 - (0.8)^n].$$

(4.4.15)

After some algebra, using (4.4.15) in (4.3.13) then gives

$$w_1(n) = 1.111 - .361(-0.8)^n - .75(0.8)^n$$

$$w_2(n) = -.389 - .361(-0.8)^n + .75(0.8)^n.$$

(4.4.16)

The dynamics of this process are shown in Figure 4.3(a), and it is seen that the convergence process is somewhat oscillatory. However, these oscillations are seen to die out, and the weights converge to

$$\lim_{N \to \infty} \mathbf{w}_N(n) = [1.111, -0.389]^T.$$

(4.4.17)

It is easy to verify by simple matrix inversion that the limit in (4.4.15) as $n \to \infty$ is $\mathbf{w}_N^*$.

To illustrate that steepest descent is rather robust with respect to the choice of α, let $\alpha = 0.5$ in (4.3.35), which gives the following solutions:

$$w_1(n) = 1.111 - .361(0.1)^n - .75(0.9)^n$$

$$w_2(n) = -.389 - .361(0.1)^n + .75(0.9)^n.$$

(4.4.18)

The convergence dynamics of this process are shown in Figure 4.3(b) in which it is seen that the weight trajectories are "smoother" than for $\alpha = 1.0$. However, it now takes longer for the weights to converge to $\mathbf{w}_N^*$. Note also that the weights do not necessarily converge monotonically toward $\mathbf{w}_N^*$. Again, whereas for solving the normal equations this slower convergence time might not be of much consequence, it is very important in applying gradient-based algorithms such as LMS to actual data. This will be more closely examined in Chapter 5.

The important point to be observed from these current examples is that the phenomena of weight oscillation, slow convergence, and nonmonotonic behavior are associated with the underlying method of steepest descent. Any gradient-based adaptive method, such as LMS or normalized LMS, will therefore exhibit this type of behavior.

4.5 Mean Square Error Propagation

Thus far, only the convergence characteristics of the steepest descent weights have been explored. However, recall from Section 4.2 that every weight vector setting, $\mathbf{w}_N(n)$, produced a corresponding point on the MSE surface that was located at a height $\varepsilon(n)$ above the $\mathbf{w}_N$-plane. It is of substantial analytical

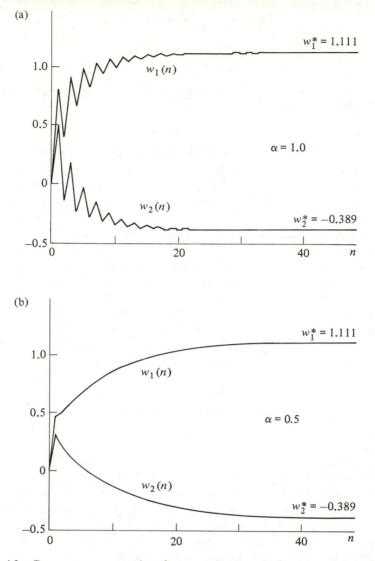

Figure 4.3 Convergence properties of steepest descent solution to normal equations (4.4.12) for two selected α values: (a) $\alpha = 1.0$, (b) $\alpha = 0.5$.

interest to examine the propagation of this MSE sequence, $\varepsilon(n)$, as it converges toward the minimum MSE value, $\varepsilon_{\min}$. The propagation of MSE is used quite frequently in the literature to quantify the performance of many adaptive methods. Since the very popular gradient-based methods are derived from the method of steepest descent, it is very beneficial to examine the MSE propagation of steepest descent. The results of this section will be the basis for examining the MSE characteristics of the LMS algorithm in Chapter 5.

It is straightforward to show that (4.2.5) is equivalent to

$$\varepsilon(n) = \varepsilon_{\min} + \mathbf{v}_N^T(n)\mathbf{R}_{NN}\mathbf{v}_N(n), \tag{4.5.1}$$

where

$$\varepsilon_{\min} = \sigma_d^2 - \mathbf{p}_N^T \mathbf{w}_N^* \tag{4.5.2}$$

is the minimum MSE achievable using the optimal filter $\mathbf{w}_N^*$, and $\mathbf{v}_N(n)$ is the weight difference vector from (2.5.3)

$$\mathbf{v}_N(n) = \mathbf{w}_N(n) - \mathbf{w}_N^*. \tag{2.5.3}$$

It should be kept in mind that (4.5.1) stills applies only to the method of steepest descent for solving the normal equations. The minimum MSE, however, is a function of the signal environment and is independent of the particular solution method chosen for the normal equations. The minimum MSE should not be associated with the residual error of the $\mathbf{w}_N(n)$ away from the optimal $\mathbf{w}_N^*$. As previously shown, steepest descent will solve the normal equations exactly for $\mathbf{w}_N^*$ as $n \to \infty$. However, depending on the particular application and signal environment, the "best" $\mathbf{w}_N^*$ might not predict $d(n)$ exactly, and this will cause a non-zero $\varepsilon_{\min}$. Examples of this were previously examined in Section 2.4. In that section, the $\mathbf{w}_N^*$ for the specific system identification problem did produce $\varepsilon_{\min} = 0$, but the $\mathbf{w}_N^*$ for the particular linear prediction problem gave $\varepsilon_{\min} \neq 0$. Similar to preceding work, (4.5.1) is unwieldy to work with directly in order to obtain theoretical results. Therefore, expand $\mathbf{R}_{NN}$ in terms of the orthogonal transformation (2.5.10), producing

$$\varepsilon(n) = \varepsilon_{\min} + \mathbf{v}_N'^T(n)\mathbf{\Lambda}_{NN}\mathbf{v}_N'(n), \tag{4.5.3}$$

where $\mathbf{v}_N'(n)$ is the uncoupled weight difference vector.

The form (4.5.3) is very compact and has a pleasing intuitive interpretation. At convergence, the gradient weight vector $\mathbf{w}_N(n)$ approaches the minimum MSE weight vector $\mathbf{w}_N^*$, and therefore $\mathbf{v}_N(n)$ approaches 0. This, in turn, implies that the uncoupled difference vector $\mathbf{v}_N'(n)$ approaches 0 as well. Therefore, the second term in (4.5.3) also approaches 0 as $n \to \infty$. Hence, the MSE in steady state for the method of steepest descent is exactly $\varepsilon_{\min}$, as required by the solution to the normal equations.

The effects of different data statistics upon the propagation of $\varepsilon(n)$ may also be examined using (4.5.3). As outlined in the problems at the end of the chapter, in may be shown that the MSE as a function of n evolves according to

$$\varepsilon(n) = \varepsilon_{\min} + \sum_{i=1}^{N} \lambda_i v_i'(0)\gamma_i^{2n}, \tag{4.5.4}$$

where $v_i'(0)$ is the initial uncoupled weight difference and $\gamma_i = 1 - \alpha\lambda_i$.

It is seen that $v_i'(0)$ is a constant once $\mathbf{w}_N(0)$ has been selected. Therefore, the convergence properties of $\varepsilon(n)$ are completely governed by the exponential behavior of γ_i^{2n}. However, using (4.3.26) it is also seen that

$$\lim_{n\to\infty} \gamma_i^n = \lim_{n\to\infty} (1 - \alpha\lambda_i)^n = 0, \tag{4.5.5}$$

provided $0 < \alpha < \lambda_{max}$. Therefore, the term γ_i^{2n} must also approach 0 if this same bound on α is met. Thus, the MSE achieved using the method of steepest descent is seen from (4.5.4) to converge exponentially to ε_{min} as n increases.

The equivalent expressions (4.5.3) and (4.5.4) will be very useful in Chapter 5 for the LMS algorithm. The LMS algorithm will be seen to approximate the method of steepest descent, and therefore the propagation characteristic of LMS will be closely related to the MSE characteristic of steepest descent.

PROBLEMS

1. Verify the linearity of the translation-and-rotation operations as applied to the method of steepest descent in the following manner:
   (a) Beginning with (4.2.10), for the method of steepest descent first perform the translation of axes to the $\mathbf{v}_N$ space, deriving a recursive equation for the $\mathbf{v}_N(n)$ coefficients. They apply the modal matrix rotation of coordinates to derive the recursion for the "rotated" (or uncoupled) $\mathbf{v}_N'(n)$ coefficients.
   (b) Now begin again with (4.2.10), but perform the modal matrix rotation first to obtain a recursion for the $\mathbf{w}_N'(n)$ coefficients. Then perform the translation of axes to derive the recursion for the $\mathbf{v}_N'(n)$ coefficients. The recursions in parts (a) and (b) should be equivalent.

2. Recall the results given in (2.3.16)–(2.3.18) for the linear prediction problem using an uncorrelated sequence $x(n)$, having mean square power σ_x^2. Solving the normal equations that resulted in this case produced $\mathbf{w}_N^* = 0$. In this problem, solve for $\mathbf{w}_N^*$ using the method of steepest descent, with $\alpha = 1/5\sigma_x^2$ and initial $\mathbf{w}_N(0) = [1, 1, \ldots, 1]^T$. Does this solution agree with $\mathbf{w}_N^*$? Explain why it does agree or why it does not agree.

3. This problem gives practice in using steepest descent to iteratively solve the matrix equation

$$\mathbf{R}_{NN}\mathbf{w}_N^* = \mathbf{p}_N$$

for $\mathbf{w}_N^*$. Let

$$\mathbf{R}_{NN} = \begin{bmatrix} 1.00 & 0.90 \\ 0.90 & 1.00 \end{bmatrix}, \qquad \mathbf{p}_N = \begin{bmatrix} 0.90 \\ 0.85 \end{bmatrix},$$

and assume that $\mathbf{w}_N(0) = [0, 0]^T$. Show that the steepest descent solution may be written as

$$\begin{bmatrix} w_1(n) \\ w_2(n) \end{bmatrix} = \begin{bmatrix} 0.71 - 0.46(1 - 1.9\alpha)^n - 0.25(1 - 0.1\alpha)^n \\ 0.21 - 0.46(1 - 1.9\alpha)^n + 0.25(1 - 0.1\alpha)^n \end{bmatrix}.$$

4. Investigate the effects of changing α in the steepest descent solution of 3 by plotting the $w_1(n)$ solution for $n = 1$ through $n = 5$ for

$$\text{(a) } \alpha = 0.1, \qquad \text{(b) } \alpha = 1.0, \qquad \text{(c) } \alpha = 3.0.$$

Why does steepest descent fail to converge for part (c)? What is the largest value of α for which steepest descent will converge?

5. Since the filter $\mathbf{w}_N(n)$ obtained by steepest descent has N coefficients that all vary, a

more common measure of the convergence characteristics of the method introduced in Section 4.5 is the MSE, $\varepsilon(n)$, as a function of n. To investigate this, compute the propagation characteristic of $\varepsilon(n)$ for steepest descent in solving the matrix equation $A_{NN}b_N = c_N$ for b_N, where:

$$A_{NN} = \begin{bmatrix} 1.0 & 0.6 \\ 0.6 & 1.0 \end{bmatrix}, \qquad c_N = \begin{bmatrix} 0.6 \\ 0.2 \end{bmatrix}.$$

Use $\alpha = 0.5$, $\sigma_d^2 = 1.0$, and $b_N^T(0) = [0, 0]$. *HINT*: Use the rotated coordinate approach to save computations.

6. Suppose it is desired to solve the matrix equation $R_{NN}w_N = p_N$ using steepest descent and the initial weights $w_N(0)$ are set equal to 0. Derive a bound on α such that the rotated uncoupled weights $v_N'(n)$ from steepest descent converge in a *monotonic* manner to their optimal values; that is, there is no "overshoot" or "oscillation" as the uncoupled weight solution $v_N'(n)$ evolves.

7. Show that (4.2.5) for the MSE propagation,

$$\varepsilon(n) = \sigma_d^2 - 2w_N^T(n)p_N + w_N^T(n)R_{NN}w_N(n)$$

is equivalent to the expression in (4.5.3).

8. Let the ith time constant of the method of steepest descent be defined as the number of iterations, τ_i, that it takes for an initial uncoupled difference weight $v_i'(0)$ to decrease to $e^{-1}v_i'(0)$ using the method of steepest descent. Assume that $v_i'(0) \neq 0$, $\tau_i \gg 1$, and $\alpha \ll 1/\lambda_{max}$. Show that the time constants may be accurately approximated by $\tau_i = 1/\alpha\lambda_i$.

REFERENCES

1. D.J. Wilde, *Optimal Seeking Methods*, Prentice-Hall, Englewood Cliffs, NJ, 1964.
2. M.R. Hestenes, *Optimization Theory*, John Wiley & Sons, New York, 1975.
3. D.A. Pierre, *Optimization Theory with Applications*, John Wiley & Sons, New York, 1969.
4. D.K. Faddeev and V.N. Faddeeva, *Computational Methods of Linear Algebra*, Freeman, San Francisco, 1963.
5. H. Robbins and S. Munro, "A Stochastic Approximation Method," *Ann. Math. Statis.*, vol. 22, pp. 400–407, 1951.
6. D.J. Sakrison, "Stochastic Approximation: A Recursive Method for Solving Difference Equations," in *Advances in Communication Theory*, vol. 2, A.V. Balakrishnan, ed., Academic Press, New York, 1966.
7. B. Widrow, "Adaptive Filters," in *Aspects of Network and System Theory*, N. de Claris and R.E. Kalman, eds., Holt, Rinehart, and Winston, New York, 1971.
8. B. Widrow and J. McCool, "A Comparison of Adaptive Algorithms Based on the Methods of Steepest Descent and Random Search," *IEEE Trans. Antennas and Propagat.*, vol. AP-24, pp. 615–637, September 1976.
9. J. Treichler, "Transient and Convergent Behavior of the ALE," *IEEE Trans. on Acous., Speech, and Signal Processing*, vol. ASSP-27, pp. 53–63, February 1979.
10. L.L. Horowitz and K.D. Senne, "Performance Advantage for Complex LMS for Controlling Narrowband Adaptive Arrays," *IEEE Trans. Acous, Speech, and Signal Processing*, vol. ASSP-29, pp. 722–736, June 1981.
11. J.A. Cadzow, "Recursive Digital Filter Synthesis Via Gradient Based Algorithms,"

IEEE Trans. Acous., Speech, and Signal Processing, vol. ASSP-24, pp. 349–355, October 1976.

12. Y. Bard, "Comparison of Gradient Methods for the Solution of Nonlinear Parameter Estimation Problems," *SIAM J. Numer. Anal.*, vol. 7, pp. 157–186, March 1970.

13. P. Eykhoff, *System Identification: Parameter and State Estimation*, John Wiley and Sons, Chichester, England, 1974.

14. B. Noble and J.W. Daniel, *Applied Linear Algebra*, Prentice-Hall, Englewood Cliffs, NJ, 1977.

15. R. Bellman, *Introduction to Matrix Analysis*, McGraw-Hill, New York, 1960.

16. G. Strang, *Linear Algebra and its Applications*, Academic Press, New York, 1976.

17. L.R. Rabiner and R.W. Schafer, *Digital Processing of Speech Signals*, Prentice-Hall, Englewood Cliffs, NJ, 1978.

18. A.V. Oppenheim and R.W. Schafer, *Digital Signal Processing*, Prentice-Hall, Englewood Cliffs, NJ, 1975.

19. J.A. Cadzow and H.F. Van Landingham, *Signals, Systems, and Transforms*, Prentice-Hall, Englewood Cliffs, NJ, 1985.

20. B.Widrow and S.D. Stearns, *Adaptive Signal Processing*, Prentice-Hall, Englewood Cliffs, NJ, 1985.

CHAPTER 5
The Least Mean Squares (LMS) Algorithm

5.1 Introduction

The method of steepest descent described and developed in the previous chapter forms the mathematical basis for many current adaptive signal processing algorithms. However, steepest descent was developed for iteratively solving the normal equations (2.3.5) for the optimal $\mathbf{w}_N^*$. For an actual signal processing application, this would be equivalent to requiring that the time series $d(n)$ and $x(n)$ be stationary and, additionally, that their second-order statistics be known. Knowledge of the second-order statistics in (2.3.5) was conveyed by the autocorrelation matrix and cross-correlation vector. However, in practical system implementations, these correlation values can only be estimated from available data, and this is often a source of computational delay, or error, or both. This chapter develops an alternative to the method of steepest descent called the least mean squares (LMS) algorithm, which will then be applied to problems in which the second-order statistics of the signal are unknown. Due to its simplicity, the LMS algorithm is perhaps the most widely used adaptive algorithm in currently implemented systems.

The LMS algorithm was introduced by Widrow [1], and since then it has been applied to a wide variety of areas. Several texts that investigate the LMS algorithm and its applications are those by Widrow and Stearns [22], Cowan and Grant [23], Honig and Messerschmitt [24], and Haykin [25]. One of the early successful applications was adaptive noise cancelling, as developed by Widrow, et al. [2], in which the linear prediction aspects of LMS were used to separate wideband information signals from narrowband interfering signals. The adaptive nature of LMS was also exploited by Widrow, et al. [3] to adaptively control the spatial response pattern of an antenna array. An extensive treatment of adaptive antenna array applications, including more advanced techniques than LMS, are contained in Monzingo and Miller [4]. In the area of spectral estimation, Griffiths [5] proposed using LMS to compute the coefficients of a rapidly adaptive all-pole spectral estimator. Later work by Treichler [6] additionally developed the concept of using LMS to enhance the sinusoidal properties of time series having low-power sinusoids in noisy backgrounds. In another application, Lucky [7] applied a gradient-based algorithm similar to LMS to the problem of reducing the intersymbol interference in data communications. Later, Sondhi [8] applied LMS to removing the echos generated on the long distance voice communications network. More recently, Alexander and Rajala [9] have applied the LMS algorithm to the problem of reducing the amount of data needed to encode images for transmission. In another communications application, Trussell and Wang [26] applied LMS as a noise canceller to decrease the error probability in a digital communications system.

Since the LMS algorithm is a stochastic adaptive algorithm, it is therefore more difficult to analyze than the deterministic method of steepest descent. Initial attempts to analyze the gradient-based adaptive methods with exact mathematical rigor often became intractable due to the presence of correlated data in the LMS filter tapped delay line. Examples of this approach are presented in the papers by Kim and Davisson [10] and Daniell [11]. Realizing this difficulty, Treichler [6] made the assumption that the filter data and filter weights were uncorrelated with each other, and derived tractable mathematical properties for the mean LMS weights during the transient adaptation period. A useful tool in this work was the modal matrix orthogonal transformation of Chapter 4, which allowed a set of uncoupled difference equations to be solved to obtain the mean LMS weight properties. This approach was then extended to a nonstationary data case for tracking the frequency of a chirped (frequency-swept) sinusoid [12]. Additionally, the LMS algorithm for complex-domain signals was advanced by Widrow, et al. in [13]. The benefit of using this configuration for modulated signal environments, as well as the significant analytical simplifications that resulted, was recognized by Bershad, et al. in [14]. This work advanced the understanding of the convergence characteristics of LMS in a very commonly occurring nonstationary signal environment.

Another property of the LMS algorithm, which will also be developed in the current chapter, is that the LMS weights have a jitter or variance associated with their instantaneous values. Initial attempts to incorporate this variance into the LMS analytical model were done by Widrow and McCool [15] for low signal-to-noise ratio environments. More recently. Horowitz and Senne [16] and Fisher and Bershad [17] proposed that the LMS weight variance is time varying and attempted to compute its value as a function of time. These resulting solutions were quite complex and, as a result, Alexander and Rajala [18] derived a simpler closed form expression for the weight variance by assuming the jitter to be a stationary process. This evolution is indicative of analysis in adaptive methods. That is, the more mathematically rigorous results are often too complex to provide easily used system information and, therefore, approximation and simplification are frequently required.

Much of this chapter is concerned with a detailed mathematical analysis of the LMS algorithm. Mathematical analysis is a very important tool, since it allows many properties of the adaptive filter to be determined without expending the time or expense of computer simulation or building actual hardware. Additionally, it is frequently quite difficult, from simulation results, to obtain the same performance insight available through analysis. Indeed, simulation is often used as a method to verify the mathematical analysis. The goal of the analysis in this chapter is to determine the LMS convergence and steady-state properties as a function of signal statistics, filter length, feedback gain, etc. The benefit is that the important parameter values become explicitly displayed through analysis, whereas rather massive sets of simulations may be necessary to isolate and quantify these parameter values. One successful previous example of this approach was seen in Chapter 4, in which analysis determined that the method of steepest descent would only converge if the feedback gain parameter α were in the range described by equation (4.4.5).

This chapter is organized as follows. Section 5.2 derives one gradient-based structure that attempts to estimate the autocorrelation function directly from the data. However, there are significant problems associated with this approach, which are discussed at length. Section 5.3 next derives the LMS algorithm by making a simple assumption in the method of steepest descent from Chapter 4. Section 5.4 then discusses the convergence of the LMS algorithm, and shows that the mean LMS weight properties are identical to those produced by the method of steepest descent, introduced in Chapter 4. Section 5.5. then concludes with an examination of the LMS mean square error (MSE) characteristics by incorporating a simple model for the LMS weight variance.

5.2 Effects of Unknown Signal Statistics

The central issue in adaptive signal processing is that there is rarely an *a priori* knowledge of signal statistics, even for truly stationary environments. In most applications, only an estimate of these signal statistics can be obtained. One approach to the problem of unknown signal statistics is to attempt to implement the method of steepest descent by estimating the gradient from the available data samples. If this approach is used, then a computational method for estimating the gradient must first be determined. One approach is to examine the expression for the true gradient, $\nabla_{\mathbf{w}}$, from (4.2.6) and formulate an estimated gradient, $\hat{\nabla}_{\mathbf{w}}$, in a similar manner. Therefore, from (4.2.6) one candidate for the form of $\hat{\nabla}_{\mathbf{w}}$ might be

$$\hat{\nabla}_{\mathbf{w}}[\varepsilon(n)] = -2\hat{\mathbf{p}}_N + 2\hat{\mathbf{R}}_{NN}\mathbf{w}_N(n), \qquad (5.2.1)$$

where the carats signify the quantities estimated from the data. The individual components of $\hat{\mathbf{R}}_{NN}$ and $\hat{\mathbf{p}}_N$ are estimates of the individual correlation function values as suggested by (2.2.6b). Estimating the individual elements of $\hat{\mathbf{R}}_{NN}$ would then produce

$$\hat{\phi}_x(m) = \frac{1}{K} \sum_{i=0}^{K-m-1} x(n-i)x(n-m-i). \tag{5.2.2}$$

The term $\hat{\phi}_x(m)$ is an estimate of the true value $\phi_x(m)$. A similar formulation gives $\hat{\phi}_{xd}(m)$, the estimate of the true cross-correlation $\phi_{xd}(m)$:

$$\hat{\phi}_{xd}(m) = \frac{1}{K} \sum_{i=0}^{K-m-1} d(n-i)x(n-m-i). \tag{5.2.3}$$

Whereas the concept of this approach is straightforward, the computational cost is substantial. A total of $K - m$ multiplications plus $K - m - 1$ additions must be computed simply to obtain each estimate, and this set of computations must be repeated for each $\hat{\phi}_x(m)$ and $\hat{\phi}_{xd}(m)$, for $0 \le m \le N$. Futhermore, in many applications such as speech and image coding, the data sequences are often nonstationary, which would require a periodic updating of the computations in (5.2.2) and (5.2.3) if this approach were used. The computational cost in this case could possibly become prohibitive for real-time applications.

Fortunately, there is a form for the gradient that is mathematically equivalent to (4.2.6), but is much simpler from a computational standpoint. Additionally, it is closely related to the orthogonality requirement discussed previously and allows the natural transition to the LMS algorithm. Recall from (2.4.1) that

$$\nabla_{\mathbf{w}}[\varepsilon(n)] = \frac{\partial}{\partial \mathbf{w}_N} E\{e^2(n)\} = 2E\{e(n)\frac{\partial}{\partial \mathbf{w}_N}e(n)\}. \tag{5.2.4}$$

By expanding $e(n)$ according to its definition in (2.2.5),

$$e(n) = d(n) - \mathbf{w}_N^T(n)\mathbf{x}_N(n), \tag{5.2.5}$$

it is immediately seen that

$$\frac{\partial}{\partial \mathbf{w}_N}e(n) = -\mathbf{x}_N(n). \tag{5.2.6}$$

Substituting this result into (5.2.4) then gives the alternate form for the gradient

$$\nabla_{\mathbf{w}}[\varepsilon(n)] = -2E\{e(n)\mathbf{x}_N(n)\},$$

and from (4.2.3), an alternate form for the steepest descent weight update becomes

$$\mathbf{w}_N(n+1) = \mathbf{w}_N(n) + \alpha E\{e(n)\mathbf{x}_N(n)\}, \tag{5.2.7}$$

where again $\alpha = 2\mu$ with no loss of generality. Therefore, this equivalent form for the steepest descent method requires that the cross-correlation of the prediction error with the signal be known. Note that if steepest descent were being applied to predicting $d(n)$ using the signal $x(n)$, then $e(n)$ and $\mathbf{x}_N(n)$, as required in (5.2.7), would be readily available. Thus, a simple computational estimate for the cross-correlation in (5.2.7) is indeed feasible and suggests the following form for the approximation to steepest descent:

$$\mathbf{w}_N(n + 1) = \mathbf{w}_N(n) + \alpha\hat{E}\{e(n)\mathbf{x}_N(n)\}, \tag{5.2.8}$$

where $\hat{E}\{\cdot\}$ signifies an estimate of the expected value.

To utilize (5.2.8) in actual computations, a computational form for $\hat{E}\{e(n)\mathbf{x}_N(n)\}$ is now needed. One form is immediately suggested by the previous example:

$$\hat{E}\{e(n)\mathbf{x}_N(n)\} = \frac{1}{K}\sum_{i=0}^{K-1} e(n - i)\mathbf{x}_N(n - i), \tag{5.2.9}$$

where K is the number of data samples used in the calculation. Note that (5.2.9) produces a vector with components given by

$$\begin{bmatrix} \hat{E}\{e(n)x(n - 1)\} \\ \hat{E}\{e(n)x(n - 2)\} \\ \cdots \\ \hat{E}\{e(n)x(n - N)\} \end{bmatrix} = \frac{1}{K} \begin{bmatrix} \sum_{i=0}^{K-1} e(n - i)x(n - i - 1) \\ \sum_{i=0}^{K-1} e(n - i)x(n - i - 2) \\ \cdots \\ \sum_{i=0}^{K-1} e(n - i)x(n - i - N) \end{bmatrix}. \tag{5.2.10}$$

The computation in (5.2.9) may be considered as operating over a window of data extending K samples into the past, and therefore incorporates only the most recent data properties. Furthermore, only N of these terms, each requiring K multiplications and $K - 1$ additions, need be calculated to estimate the entire gradient vector. Substituting (5.2.9) into (5.2.8) then gives

$$\mathbf{w}_N(n + 1) = \mathbf{w}_N(n) + \frac{\alpha}{K}\sum_{i=0}^{K-1} e(n - i)x_N(n - i). \tag{5.2.11}$$

Equation (5.2.11) thus provides one method of estimating the gradient and gives an adaptive implementation for updating the $\mathbf{w}_N(n)$ coefficients directly from the data. However, the amount of computation involved in (5.2.10) is still substantial, and the smoothing effect of the K-sample time window is significant. Therefore, the method of (5.2.11) has not found a great deal of application in practical problems. Instead, a very simple approximation to the gradient in (5.2.7) may be used, leading to the LMS algorithm that is the topic of the next section.

5.3 Derivation of the LMS Algorithm

A very useful algorithm has evolved from simply approximating the expectation in (5.2.7) with the instantaneous value of the quantity inside the brackets. That is, let the estimate of the expected value be given simply as

$$\hat{E}\{e(n)\mathbf{x}_N(n)\} = e(n)\mathbf{x}_N(n). \tag{5.3.1}$$

Substitution of (5.3.1) into (5.2.8) then leads directly to the LMS algorithm:

$$\mathbf{w}_N(n + 1) = \mathbf{w}_N(n) + \alpha e(n)\mathbf{x}_N(n). \tag{5.3.2}$$

Due mainly to its simplicity, the LMS algorithm has found wide usage in applications that deal with nonstationary data or time-varying statistics.

The LMS algorithm is sometimes referred to as the noisy gradient or gradient approximation algorithm, since the structure of the method of steepest descent can be preserved by defining the noisy gradient as the gradient of the instantaneous squared error (rather than the gradient of the expected squared error):

$$\hat{\nabla}_\mathbf{w}[\varepsilon(n)] = \frac{\partial}{\partial \mathbf{w}_N} e^2(n) = -2e(n)\mathbf{x}_N(n). \tag{5.3.3}$$

It follows from substitution of (5.3.3) into (5.3.2) that the LMS algorithm may be equivalently written as

$$\mathbf{w}_N(n + 1) = \mathbf{w}_N(n) - \mu\hat{\nabla}_\mathbf{w}[\varepsilon(n)], \tag{5.3.4}$$

where $\mu = \alpha/2$.

The popularity of the LMS algorithm stems largely from the simplicity of its computational structure, low storage requirements, and the relative ease with which it may be mathematically analyzed. In this chapter, the computational and analytical properties of the LMS algorithm will be examined. It will be seen that many of the analytical results of Chapter 4 for the method of steepest descent will hold for LMS, which should not be surprising since the LMS algorithm has been seen to be an approximation to the method of steepest descent.

Suppose (5.3.2) is expanded in terms of each vector component. Then it is easy to derive the scalar update for each weight:

$$w_i(n + 1) = w_i(n) + \alpha e(n)x(n - i); \qquad 1 \le i \le N. \tag{5.3.5}$$

Equation (5.3.5) explicitly displays one very important difference between LMS and the method of steepest descent. In Chapter 4, it was found that the statistical property of orthogonality between the prediction error and the filter data must hold for the method of steepest descent to converge. However, in (5.3.5), for the LMS algorithm, there is a much more stringent condition for the $w_i(n)$ to converge exactly to w_i^*, namely,

$$e(n)x(n - i) = 0; \qquad 1 \le i \le N,$$

which is obtained simply by enforcing that $w_i(n + 1) = w_i(n)$ in (5.3.5). This condition must hold *for every time iteration n* in order for the $\mathbf{w}_N(n)$ produced by LMS to converge to the optimal $\mathbf{w}_N^*$. In general, this condition is quite difficult to achieve in practice since it requries that every product $e(n)x(n - i)$ must be zero, and not just that the expectation of this product vanish. Therefore, an initial impression concerning LMS is that compared to steepest descent, the LMS steady-state solutions might be "suboptimal" in some way. It will soon be seen that at convergence, the instantaneous values of the LMS weights are not exactly equal to the optimal values; however, the expected value or mean of the LMS weight vector $\mathbf{w}_N(n)$ will be seen to converge to the optimal $\mathbf{w}_N^*$ as $n \to \infty$, which is the same result derived in Chapter 4 for the method of steepest descent.

5.4 Convergence of the LMS Algorithm

In this section, some of the convergence properties of the LMS algorithm will be examined, which will draw heavily from previous results for the method of steepest descent from Chapter 4. An emphasis will be placed upon when the previous results for steepest descent will be directly applicable, and when the stochastic nature of the LMS algorithm requires the development of new analytical results.

This study of convergence of the LMS algorithm will investigate properties of the LMS adaptive prediction filter coefficients as they adapt from a set of initial conditions to a final set of converged coefficients. Begin with (5.3.2) for the LMS weight vector update,

$$\mathbf{w}_N(n + 1) = \mathbf{w}_N(n) + \alpha e(n)\mathbf{x}_N(n), \tag{5.4.1}$$

and the definition of the LMS prediction error

$$e(n) = d(n) - \mathbf{x}_N^T(n)\mathbf{w}_N(n). \tag{5.4.2}$$

Substitution of (5.4.2) into (5.4.1) then gives

$$\mathbf{w}_N(n + 1) = [\mathbf{I}_{NN} - \alpha\mathbf{x}_N(n)\mathbf{x}_N^T(n)]\mathbf{w}_N(n) + \alpha d(n)\mathbf{x}_N(n). \tag{5.4.3}$$

Equation (5.4.3) may be compared directly to (4.2.10) for the method of steepest descent. Rewriting (4.2.10) for convenience,

$$\mathbf{w}_N(n + 1) = [\mathbf{I}_{NN} - \alpha\mathbf{R}_{NN}]\mathbf{w}_N(n) + \alpha\mathbf{p}_N, \tag{4.2.10}$$

it is seen that (5.4.3) for the LMS algorithm would correspond directly to (4.2.10) if the matrix $\mathbf{x}_N(n)\mathbf{x}_N^T(n)$ were replaced by the $\mathbf{R}_{NN}$ matrix, and if the vector $d(n)\mathbf{x}_N(n)$ were replaced by the $\mathbf{p}_N$ vector. However, recall from (2.2.8) and (2.2.12) the definitions of $\mathbf{R}_{NN}$ and $\mathbf{p}_N$:

$$\mathbf{R}_{NN} = E\{\mathbf{x}_N(n)\mathbf{x}_N^T(n)\} \tag{2.2.12}$$

$$\mathbf{p}_N = E\{d(n)\mathbf{x}_N(n)\}. \tag{2.2.8}$$

It is immediately seen that steepest descent utilizes the expectations of the instantaneous quantities displayed in the LMS algorithm.

This similarity is to be expected, since LMS approximates the method of steepest descent. The primary difference between steepest descent and LMS is that steepest descent is deterministic and requires the exact gradient, whereas LMS is a stochasic recursive algorithm and uses only a noisy approximation of the gradient. In some of the literature, LMS and its variants are even referred to as stochastic gradient or stochastic approximation methods. The terminology *stochastic recursion* signifies that the update term, or the driving function of the stochastic difference equation (5.3.2) is a member of a stochastic process. This is true since, in LMS, the sequences $x(n)$ and $d(n)$ are members of stochastic processes. However, the method of steepest descent from (4.2.10) is *deterministic*; that is, the parameters $\mathbf{R}_{NN}$ and $\mathbf{p}_N$ are assumed to be known and constant.

The LMS algorithm, on the other hand, uses only the acquired signals to calculate its recursion for the LMS weight update. If the signals $x(n)$ and $d(n)$ were deterministic, then the LMS algorithm would likewise be deterministic, provided the initial $\mathbf{w}_N(0)$ were constant for all trials. However, when $x(n)$ becomes a stochastic signal governed by a probability density, then the situation becomes more complicated since in theory, at least, the specific values of $x(n)$ are never known exactly. A common example of a stochastic signal might be a known sinusoid corrupted by additive, uncorrelated random noise.

One common approach to analyzing stochastic adaptive systems is to partition the problem into the considerations of solving separately for the mean weight behavior and the weight variance characteristics. The LMS predictor weights will have some general probability density function (usually unknown, however) from which the mean (i.e., expected value) and the variance of the weights could theoretically be computed. Unfortunately, the computation of this probability density is exceedingly difficult, if not impossible, and is well beyond the scope of the present text. For an example of this type of approach, the interested reader is referred to the works by Berni [19], Fisher and Bershad [17], Bershad and Qu [20], and Iltis and Milstein [21]. Engineering analysis is often the science of judicious assumption, and a great deal of insight may be derived by separate considerations of the LMS weight expectation and weight variance. By employing some typically nonrestrictive assumptions concerning the LMS weights and signal data, some very tractable and useful analytical results may be derived. It will be shown that the assumptions used in this approach are valid for a large number of signal processing applications.

The mean LMS weight vector

To examine the behavior of the mean value of the LMS weight vector, simply take the expectation of both sides of (5.4.3):

$$E\{\mathbf{w}_N(n+1)\} = E\{[\mathbf{I}_{NN} - \alpha\mathbf{x}_N(n)\mathbf{x}_N^T(n)]\mathbf{w}_N(n)\} + \alpha E\{d(n)\mathbf{x}_N(n)\}$$
$$= E\{\mathbf{w}_N(n)\} - \alpha E\{[\mathbf{x}_N(n)\mathbf{x}_N^T(n)]\mathbf{w}_N(n)\} + \alpha E\{d(n)\mathbf{x}_N(n)\}.$$
$$(5.4.4)$$

The second expectation in (5.4.4) can be simplified by using the following assumption:

Assumption 1. *The data signal $x(n)$ and the LMS weights, $w_i(n)$, are uncorrelated with each other.*

In general, the conditions in Assumption 1 are fulfilled by data $x(n)$ and weights $w_i(n)$ that are sufficiently "dissimilar" over the time of observation. One commonly occurring environment in which this assumption holds very well is when the signal changes much more rapidly than the mean value of the weights. For example, a speech waveform might change very rapidly, whereas the LMS weights will attempt to adapt to the statistical model of the speech, which may be fairly constant over the observation time. Usually, this implies the feedback gain parameter, α, of the LMS filter should be fairly small, such that the update term in (5.3.2) adds only a small increment to the previous weight value. This discussion is meant to illustrate that whereas a rigorous proof of Assumption 1 would probably be very difficult, a qualitative justification of its viability may often be obtained from consideration of the physical situation. For the wide majority of practical applications, the model of (5.4.4) will indeed be valid, and the error introduced by making this assumption is often negligible.

Therefore, incorporating Assumption 1 allows (5.4.4) to be written as

$$E\{\mathbf{w}_N(n+1)\} = [\mathbf{I}_{NN} - \alpha\mathbf{R}_{NN}]E\{\mathbf{w}_N(n)\} + \alpha\mathbf{p}_N, \qquad (5.4.5)$$

which is exactly (4.2.10) with $\mathbf{w}_N(n)$ from the steepest descent algorithm replaced by the mean value of the LMS weight vector, $E\{\mathbf{w}_N(n)\}$. Therefore, the mean LMS weight vector propagates according to the same recursion as the weight vector obtained by the method of steepest descent. It is necessary to keep in mind that the applicability of (5.4.5) is subject to the conditions outlined in Assumption 1. The notation and analysis of this chapter will be simplified a great deal if the mean LMS weight vector, $\tilde{\mathbf{w}}_N(n)$, is defined as the expectation of the instantaneous weight vector:

$$\tilde{\mathbf{w}}_N(n) = E\{\mathbf{w}_N(n)\}. \qquad (5.4.6)$$

The individual components are therefore given by

$$\tilde{w}_i(n) = E\{w_i(n)\}. \qquad (5.4.7)$$

With this notation, (5.4.5) simplifies to

$$\tilde{\mathbf{w}}_N(n+1) = [\mathbf{I}_{NN} - \alpha\mathbf{R}_{NN}]\tilde{\mathbf{w}}_N(n) + \alpha\mathbf{p}_N. \qquad (5.4.8)$$

Note that this result is equivalent to (4.2.10) for the method of steepest descent, with the $\mathbf{w}_N(n)$ of steepest descent replaced by the mean LMS weight vector

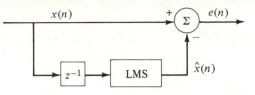

Figure 5.1   LMS algorithm used as a linear predictor.

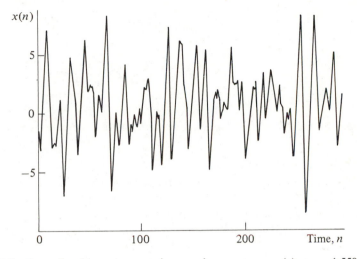

Figure 5.2   Second-order autoregressive random process $x(n)$; $a_1 = 1.558$, $a_2 = -0.81$, $\sigma_v^2 = 1.0$.

$\tilde{\mathbf{w}}_N(n)$. The immediate impact is that (4.3.35) may be used to compute the trajectories of the mean LMS weights.

As an example of the properties of the expected weight vector, consider Figure 5.1, which illustrates the LMS algorithm used in a linear prediction application. The input signal $x(n)$ is shown in Figure 5.2 and was created by a second-order autoregressive (AR) process having poles at $(z_1, z_2) = 0.9e^{\pm j\pi/6}$. It is easy to show that this AR process leads to the following difference equation for the second-order system:

$$x(n) = 1.558x(n - 1) - 0.81x(n - 2) + v(n), \tag{5.4.9}$$

where $v(n)$ is a zero mean, uncorrelated stochastic excitation sequence. A two coefficient LMS filter was used to predict $x(n)$ by

$$\hat{x}(n) = \sum_{i=1}^{2} w_i(n)x(n - i + 1). \tag{5.4.10}$$

Since, in this example, there are two AR coefficients in the signal model and there are two coefficients in the LMS filter, then $w_1(n)$ should converge

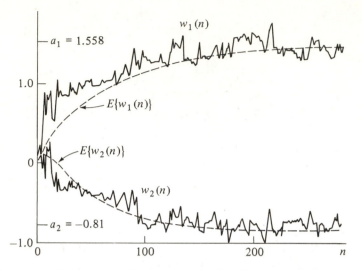

Figure 5.3 Convergence characteristics for LMS linear predictor, $\alpha = .02$.

to 1.558 and $w_2(n)$ should converge to -0.81. Figure 5.3 shows the convergence characteristics of the weights for a choice of $\alpha = 0.02$. There is seen to
be a great deal of fluctuation in the values of the weights as they converge to
a neighborhood around their optimal values a_1 and a_2. This is due to the fact
that the "noisy" gradient estimate is used in the LMS algorithm, rather than
the true gradient (which is unknown!). For comparison, the mean weight
trajectories computed from (5.4.8) for this choice of α are also shown in
Figure 5.3.

Equation (4.3.35) for the convergence characteristics of the mean LMS
weights also shows that the mean weights should take longer to converge
if α is decreased. Figure 5.4 illustrates this phenomenon by showing the
convergence results of one trial using $\alpha = 0.004$. Comparing these results with
Figure 5.3, it is indeed seen that the LMS weights take longer to converge for
the smaller α. However, there is another property evident in Figure 5.3;
namely, that decreasing α causes the weight trajectory to be much "smoother,"
or have less variance about the mean value. This is a fundamental tradeoff in
using the LMS algorithm: high α values cause the mean weights to converge
more quickly, but also cause the instantaneous weights to fluctuate more. The
appropriate α selection will depend upon the degree of accuracy needed and
the convergence speed required by the particular application.

The importance of the analysis of this chapter is that the convergence
results of Chapter 4 for the steepest descent weights apply immediately to the
convergence properties of the mean LMS weights. No new analysis need be
done to derive the mean LMS convergence or steady-state weight solutions.
However, as illustrated in Figures 5.3 and 5.4, since the instantaneous LMS
weights "jitter" about their mean value, this causes an instantaneous error

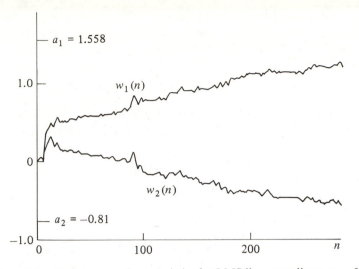

Figure 5.4   Convergence characteristics for LMS linear predictor, $\alpha = .004$.

in the LMS weight values, and therefore an excess in the prediction error compared to that suggested by simply using the LMS mean weight equations. The degree of this excess error can be quantified by analysis, which is done by deriving the propagation characteristics for the LMS mean square error (MSE). This is the subject of the next section.

5.5 LMS Mean Square Error Propagation

In Section 5.4, solutions were developed for the mean of the LMS weight vector. The actual weights themselves fluctuate about this mean level $\tilde{\mathbf{w}}_N(n)$ since LMS only approximates the true gradient with the instantaneous product of (5.3.3). A common method of modeling this weight fluctuation is to assume that the actual LMS weight vector, $\mathbf{w}_N(n)$, is the sum of the mean LMS weight vector, $\tilde{\mathbf{w}}_N(n)$, plus a stochastic (i.e., random) noisy component, $\mathbf{z}_N(n)$, where

$$\mathbf{w}_N(n) = \tilde{\mathbf{w}}_N(n) + \mathbf{z}_N(n). \tag{5.5.1}$$

This weight fluctuation affects the performance of the LMS prediction filter since it causes the actual LMS weight to be different from the value that would result from steepest descent. This effectively increases the mean square prediction error obtained using LMS over that obtained using steepest descent. However, it must be emphasized that steepest descent is rarely applicable in practical situations, since the signal statistics are rarely known *a priori*, and therefore are unavailable for use in steepest descent. Therefore, it is of substantial interest to calculate the mean square error (MSE) propagation characteristic of the LMS algorithm, since this MSE characteristic is often the measure for judging performance of the different adaptive methods.

Consistent with previous developments, define the LMS mean square prediction error as

$$\varepsilon(n) = E\{e^2(n)\} = E\{[d(n) - \mathbf{w}_N^T(n)\mathbf{x}_N(n)]^2\}, \qquad (5.5.2)$$

where the LMS prediction error from (5.2.5) has been used. Performing the operations in (5.5.2) and using the linearity of the expectation operation then gives

$$\begin{aligned}
\varepsilon(n) = E\{d^2(n)\} &- 2E\{\mathbf{w}_N^T(n)\mathbf{x}_N(n)d(n)\} \\
&+ E\{\mathbf{w}_N^T(n)\mathbf{x}_N(n)\mathbf{x}_N^T(n)\mathbf{w}_N(n)\}.
\end{aligned} \qquad (5.5.3)$$

Then substituting (5.5.1) for the LMS weights into (5.5.3) produces

$$\begin{aligned}
\varepsilon(n) = E\{d^2(n)\} &- 2E\{\tilde{\mathbf{w}}_N^T(n)\mathbf{x}_N(n)d(n)\} + E\{\tilde{\mathbf{w}}_N^T(n)\mathbf{x}_N(n)\mathbf{x}_N^T(n)\tilde{\mathbf{w}}_N(n)\} \\
&- 2E\{\mathbf{z}_N^T(n)\mathbf{x}_N(n)d(n)\} + E\{\mathbf{z}_N^T(n)\mathbf{x}_N(n)\mathbf{x}_N^T(n)\tilde{\mathbf{w}}_N(n)\} \\
&+ E\{\tilde{\mathbf{w}}_N^T(n)\mathbf{x}_N(n)\mathbf{x}_N^T(n)\mathbf{z}_N(n)\} + E\{\mathbf{z}_N^T(n)\mathbf{x}_N(n)\mathbf{x}_N^T(n)\mathbf{z}_N(n)\}.
\end{aligned} \qquad (5.5.4)$$

In general, (5.5.4) is rather imposing. However, by making use of some previously derived results and the properties of the weight noise model (5.5.1), the weight noise expression may be simplified substantially. First, however, consider the first three terms on the right-hand side of (5.5.4), which are seen to be entirely a function of the mean LMS weights $\tilde{\mathbf{w}}_N(n)$. Since it is known that the mean LMS weights are equivalent to the weights from steepest descent, the results from Section 4.5 immediately apply to these terms. From (4.5.3), the first three terms on the right-hand side of (5.5.4) are seen to be the MSE given by steepest descent at iteration n. Therefore, incorporating the result (4.5.3) into (5.5.4) gives

$$\begin{aligned}
\varepsilon(n) = \varepsilon_{\min} &+ \mathbf{v}_N'^T(n)\boldsymbol{\Lambda}_{NN}\mathbf{v}_N'(n) \\
&- 2E\{\mathbf{z}_N^T(n)\mathbf{x}_N(n)d(n)\} + E\{\mathbf{z}_N^T(n)\mathbf{x}_N(n)\mathbf{x}_N^T(n)\tilde{\mathbf{w}}_N(n)\} \\
&+ E\{\tilde{\mathbf{w}}_N^T(n)\mathbf{x}_N(n)\mathbf{x}_N^T(n)\mathbf{z}_N(n)\} + E\{\mathbf{z}_N^T(n)\mathbf{x}_N(n)\mathbf{x}_N^T(n)\mathbf{z}_N(n)\}.
\end{aligned} \qquad (5.5.5)$$

The value $\varepsilon_{\min}$ is the minimum MSE achievable by the optimal filter $\mathbf{w}_N^*$, $\mathbf{v}_N'(n)$ is the uncoupled difference weight vector, and $\boldsymbol{\Lambda}_{NN}$ is the diagonal matrix of eigenvalues of $\mathbf{R}_{NN}$, all exactly as previously discussed in Chapter 4.

It remains to compute the MSE contribution from the last four terms of (5.5.4), which is the increase in MSE due to using LMS. To help simplify this problem, the properties of the LMS weight noise model from (5.5.1) may be used. It should be emphasized that the true probability density of the $z_i(n)$ noise terms is, in general, unknown and can really only be approximated according to a model. The model chosen in this section leads to a mathematically tractable analysis, and is consistent with experimental results for a wide range of applications. In the notation to follow, $\delta(k)$ is the unit sample function, defined as

$$\delta(k) = \begin{cases} 1, & k = 0 \\ 0, & else. \end{cases}$$

These assumed properties of the weight noise model are listed as follows:

Assumption 2. *The weight noise components, $z_i(n)$, are zero mean:*

$$E\{z_i(n)\} = 0. \tag{5.5.6a}$$

Assumption 3. *The weight noise components are mutually uncorrelated, with mean square power σ_z^2:*

$$E\{z_i(n)z_j(n)\} = \sigma_z^2 \delta(i - j). \tag{5.5.6b}$$

Assumption 4. *The $z_i(n)$ are uncorrelated with respect to time:*

$$E\{z_i(n - n_1)z_i(n - n_2)\} = 0. \tag{5.5.6c}$$

Assumption 5. *The $z_i(n)$ are uncorrelated with the mean LMS weight components:*

$$E\{z_i(n)\tilde{w}_j(n)\} = E\{z_i(n)\}E\{\tilde{w}_j(n)\} = 0. \tag{5.5.6d}$$

Assumption 6. *The $z_i(n)$ are uncorrelated with the data signal*

$$E\{z_i(n)x(n)\} = E\{z_i(n)\}E\{x(n)\} = 0. \tag{5.5.6e}$$

Assumptions 2–6 greatly simplify three of the remaining four terms in (5.5.4). Indeed, it is straightforward to show that the enforcement of conditions (5.5.6a)–(5.5.6e) on the weight noise model leads to the following results:

$$E\{\mathbf{z}_N^T(n)\mathbf{x}_N(n)\} = 0 \tag{5.5.7}$$

$$E\{\mathbf{z}_N^T(n)\mathbf{x}_N(n)\mathbf{x}_N^T(n)\tilde{\mathbf{w}}_N(n)\} = E\{\tilde{\mathbf{w}}_N^T(n)\mathbf{x}_N(n)\mathbf{x}_N^T(n)\mathbf{z}_N(n)\} = 0. \tag{5.5.8}$$

Therefore, three of the terms in (5.5.4) vanish as a result of the properties of the weight noise model of (5.5.6). Therefore, the only remaining calculation is to find the MSE contribution of the last term in (5.5.4). This term is often referred to as the "excess MSE" due to the stochastic fluctuation of the LMS weights about the mean values.

Excess mean square error

A rigorous and exact calculation of a closed form expression for the excess MSE contribution would require a knowledge of the fourth-order joint probability density functions of the LMS weights. However, this approach is extremely involved and is well beyond the scope of this text. A simpler approach utilizes Assumptions 2–6, which are usually valid for the present applications and result in a substantial simplification in the analysis. To simplify the derivation, use the property (5.5.6e) that the weight noise $\mathbf{z}_N(n)$ is uncorrelated with the data $\mathbf{x}_N(n)$. Performing the expectation required by the last term in (5.5.4) then gives

$$E\{\mathbf{z}_N^T(n)\mathbf{x}_N(n)\mathbf{x}_N^T(n)\mathbf{z}_N(n)\} = \sigma_x^2 \, Tr[E\{\mathbf{z}_N(n)\mathbf{z}_N^T(n)\}], \qquad (5.5.9)$$

where $Tr[\mathbf{A}]$ signifies the trace of the matrix $\mathbf{A}$. The result in (5.5.9) may be verified by simply expanding the left-hand side of (5.5.9) and then invoking the uncorrelated properties of the weight noise model from (5.5.6). Combining the results (5.5.5) through (5.5.9), the MSE propagation for the LMS algorithm now simplifies to

$$\varepsilon(n) = \varepsilon_{\min} + \mathbf{v}_N'^T(n)\mathbf{\Lambda}_{NN}\mathbf{v}_N'(n) + \sigma_x^2 \, Tr[E\{\mathbf{z}_N(n)\mathbf{z}_N^T(n)\}]. \qquad (5.5.10)$$

The matrix $E\{\mathbf{z}_N(n)\mathbf{z}_N^T(n)\}$ is called the LMS weight covariance matrix, and it specifies the second-order statistics of the LMS weight fluctuations about their mean values. The rigorous computation of the LMS weight covariance matrix is once again a complex problem well beyond the scope of current applications, and is the subject of much present research in adaptive signal processing [19, 20]. For present purposes, an approximation will be used that is equivalent to an assumption that the weight noise $\mathbf{z}_N(n)$ is a stationary signal. This basically means that the power of $\mathbf{z}_N(n)$ is constant, regardless of whether the LMS algorithm is in the transient converging mode (where n is small) or in the steady-state converged mode ($n \to \infty$). While this assumption is not absolutely true in all cases, its validity is stronger as the feedback gain α becomes smaller. For the wide majority of the applications considered in this book, α is indeed in this required region such that the stationarity of the weight noise is a valid assumption.

The remaining computations for the weight covariance matrix and the necessary approximations are explored in the problems at the end of the chapter. The final result shows that

$$Tr[E\{\mathbf{z}_N(n)\mathbf{z}_N^T(n)\}] \approx \sum_{i=1}^{N} \frac{\alpha\varepsilon_{\min}}{(2 - \alpha\lambda_i)}. \qquad (5.5.11)$$

Therefore, incorporating the results of (5.5.11) into (5.5.10) produces the useful expression for the MSE propagation as a function of time n for the LMS algorithm:

$$\varepsilon(n) = \varepsilon_{\min} + \mathbf{v}_N'^T(n)\mathbf{\Lambda}_{NN}\mathbf{v}_N'(n) + \sigma_x^2\alpha\varepsilon_{\min}\sum_{i=1}^{N}(2 - \alpha\lambda_i)^{-1}. \qquad (5.5.12)$$

Note that the first two terms on the right-hand side of (5.5.12) are equivalent to (4.5.3), which gives the MSE incurred using the method of steepest descent. Since it has been shown that the analysis of the steepest descent weight vector corresponds to the analysis of the mean LMS weight vector, then these first two terms may be considered as the MSE contribution due to the mean LMS weight vector $\tilde{\mathbf{w}}_N(n)$. Therefore, the last term on the right-hand side of (5.5.12) is due to the actual LMS weight jitter, and this contribution is sometimes referred to as the *excess MSE* or the *misadjustment noise*. The relative power of this misadjustment noise is dependent on both data and filter parameters, as seen by (5.5.12). Some of the properties of this misadjustment noise are investigated in the problems at the end of the chapter.

The LMS algorithm has been widely applied in many areas of communications and signal processing. Several of these are illustrated in Chapter 6 [and for additional approaches to LMS the reader is directed to 4,22–25].

PROBLEMS

1. Suppose the $N = 2$ coefficient LMS algorithm is operating in the prediction mode and has been converged for a long time at time $n = n_0$ for an input signal having autocorrelation values

$$\phi_x(0) = 1.00, \qquad \phi_x(1) = 0.90, \qquad \phi_x(2) = 0.70.$$

At time $n = n_0$, let the data undergo a change in statistics such that the auto-correlation coefficients become

$$\phi_x(0) = 1.00, \qquad \phi_x(1) = 0.50, \qquad \phi_x(2) = 0.20.$$

(a) What are the initial mean LMS weights at $n = n_0$? What bound on α is required to produce this mean weight vector?

(b) What are the converged mean LMS weights for $n \to \infty$? What bounds on α are required to produce this mean weight vector?

(c) Draw the mean LMS weight values in the transition region for $n = n_0, n_0 \pm 10, n_0 \pm 20, n_0 \pm 30$. Use $\alpha = 0.2$.

2. As discussed in Chapter 4 for the steepest descent method, the MSE, $\varepsilon(n)$, for the LMS algorithm is a function of time and often displays the convergence properties of adaptive filters more succintly than do the weights themselves. P1 above will be reexamined using this approach. Assume the filter has converged long before $n = n_0$.

(a) For $\alpha = 0.2$, plot the LMS mean square prediction error from $n = n_0 - 30$ to $n_0 + 30$ in increments of $n = 10$. Disregard the misadjustment error due to LMS weight variance in this case. Note that this is equivalent to the performance of the method of steepest descent.

(b) Now include the effects of the weight jitter or variance that produces the misadjustment noise. Plot the LMS MSE for the same time values as in the (a) part.

3. The main difference between the weights produced by the method of steepest descent and the LMS weights is that the LMS weights vary about their mean values. This effect is often modeled as a random noise component that interferes with the prediction of $d(n)$ and passes an excess error through to the prediction error sequence. Since this excess MSE component is due to the misadjustment of the LMS weights away from the mean values, it is often called and misadjustment noise.

(a) Assume the algorithm has converged and find the excess MSE incurred from using the $N = 1$ and $N = 2$ coefficient LMS algorithm as a prediction filter in the following environment:

$$\phi_x(0) = 1.00, \qquad \phi_x(1) = 0.90, \qquad \phi_{xd}(2) = 0.70,$$

$$\alpha = 0.2.$$

(b) Now find the total MSE incurred from using each LMS filter in part (a).

4. Another parameter that dramatically influences the LMS misadjustment noise is the feedback gain parameter α. For the two coefficient LMS filter used as a linear predictor, calculate and plot the misadjustment noise (at convergence) for $\alpha = 1.0$, $0.5, 0.1, 0.05$, and 0.01 in each of the following signal environments:
(a) $\phi_x(0) = 1.0, \qquad \phi_x(1) = 0.9, \qquad \phi_x(2) = 0.7.$
(b) $\phi_x(0) = 1.0, \qquad \phi_x(1) = 0.5, \qquad \phi_x(2) = 0.2.$

5. Assume that the LMS algorithm is used in the systems identification application shown in Figure 1.1. A zero mean, white noise signal is input to the system with impulse response $h(n) = a\delta(n - \tau)$, where a is the system attenuation and τ is an integer representing the time delay of the system.
(a) Find the steady-state, length N, mean LMS weight vector, assuming $N < \tau$.
(b) Find the steady-state, length N, mean LMS weight vector assuming $N \geq \tau$.
(c) Now assume $N > \tau$ and the initial weight vector $\mathbf{w}_N(0) = 0$. Find the mean LMS weight trajectories for all N weights for $n \geq 0$.
(d) Rework part (c) assuming $w_i(0) = 1$ for all i. Assume that $v_i(0) \neq 0$, $\tau_i \gg 1$, and $\alpha \ll 1/\lambda_{\max}$. Show that the time constants may be accurately approximated by $\tau_i = 1/\alpha\lambda_i$.

6. Let $x(n)$ be a zero mean signal with mean square value σ_x^2. Consider the case in which a variant of the LMS algorithm predictor is used as a linear predictor. The linear prediction error is

$$e(n) = x(n) - \sum_{k=1}^{N} w_k(n)x(n - k),$$

and the predictor weights are updated according to

$$w_k(n + 1) = \beta w_k(n) + \alpha e(n)x(n - k),$$

where $\beta < 1$. This is sometimes called the "leaky" LMS algorithm. Find the solution for $E\{\mathbf{w}_N(n)\}$, where $\mathbf{w}_N(n)$ is the vector of the $w_k(n)$ at time n. What are the bounds on α for stability? Assume $\mathbf{w}_N(0) = 0$ and that $\mathbf{w}_N(n)$ and $x(n)$ are statistically independent. What is an application in which the leaky LMS algorithm might be of benefit?

7. Suppose a causal linear system h_k has input $x(n)$ and output $d(n)$. Find the values of the h_k if the following is known:

$$\phi_x(m) = \delta(m), \qquad \phi_{xd}(0) = 0.8, \qquad \phi_{xd}(1) = 0.5,$$

$$\phi_{xd}(m) = 0, \qquad \text{all other } m.$$

8. Find the functional form for the mean LMS weight trajectories if the LMS algorithm is used to find the unknown system in P7. Use $N = 2$, $\mathbf{w}_N(0) = 0$, and find $w_1(n)$ and $w_2(n)$. Do they converge to the unknown system in P7? Why or why not?

9. Suppose the LMS predictor is used to solve a systems identification problem in which $x(n)$ is the systems input and $d(n)$ is the systems output. Let $x(n)$ be

uncorrelated, zero mean, with variance σ_x^2, and let the system h_k be given by

$$h_k = \begin{cases} a, & k = 0 \\ b, & k = 1 \\ 0, & else, \end{cases}$$

and let the predicted output be

$$\hat{d}(n) = w_1(n)x(n) + w_2(n)x(n-1).$$

Find the functional form for $E\{\mathbf{w}_N(n)\}$ as a function of a and b.

REFERENCES

1. B. Widrow, "Adaptive Filters," in *Aspects of Network and System Theory*, N. de Claris and R.E. Kalman, eds., Holt, Rinehart, and Winston, New York, 1971.
2. B. Widrow, et al., "Adaptive Noise Cancelling: Principles and Applications," *Proceedings of the IEEE*, vol. 63, pp. 1962–1976, December 1975.
3. B. Widrow, et al., "Adaptive Antenna Systems," *Proceedings of the IEEE*, vol. 55, pp. 2143–2159, December 1967.
4. R.A. Monzingo and T.W. Miller, *Introduction to Adaptive Arrays*, Wiley-Interscience, New York, 1980.
5. L.J. Griffiths, "Rapid Measurement of Digital Instantaneous Frequency," *IEEE Trans. on Acous., Speech and Signal Processing*, vol. ASSP-22, pp. 207–222, April 1975.
6. J. Treichler, "Transient and Convergent Behavior of the ALE," *IEEE Trans. on Acous., Speech and Signal Processing*, vol. ASSP-27, pp. 53–63, February 1979.
7. R.W. Lucky, "Techniques for Adaptive Equalization of Digital Communications Systems," *Bell Sys. Tech. J.*, vol. 45, no. 2, 1966.
8. M.M. Sondhi, "An Adaptive Echo Canceller," *Bell Sys. Tech. J.*, vol. 46, no. 3, 1967.
9. S.T. Alexander and S.A. Rajala, "Image Compression Results Using the LMS Adaptive Algorithm," *IEEE Trans. on Acous., Speech, and Signal Processing*, vol. ASSP-33, pp. 712–715, June 1985.
10. J. Kim and L. Davisson, "Adaptive Linear Estimation for Stationary M-Dependent Processes," *IEEE Trans. on Information Theory*, vol. IT-21, pp. 23–31, January 1975.
11. T. Daniell, "Adaptive Estimation with Mutually Correlated Training Samples," *IEEE Trans. Sys. Sci. Cybernetics*, vol. SSC-6, pp. 12–19, January 1970.
12. J. Treichler, "Response of the Adaptive Line Enhancer to Chirped and Doppler Shifted Sinusoids," *IEEE Trans. on Acous., Speech and Signal Processing*, vol. ASSP-28, pp. 343–348, June 1980.
13. B. Widrow, J. McCool, and M. Ball, "The Complex LMS Algorithm," *Proceedings of the IEEE*, vol. 63, pp. 719–720, April 1975.
14. N. Bershad, et al., "Tracking Characteristics of the LMS Adaptive Line Enhancer Response to a Linear Chirp Signal in Noise," *IEEE Trans. on Acous. Speech, and Signal Processing*, vol. ASSP-28, pp. 504–516, October 1980.
15. B. Widrow and J. McCool, "A Comparison of Adaptive Algorithms Based on the Methods of Steepest Descent and Random Search," *IEEE Trans. Antennas and Propagat.*, vol. AP-24, pp. 615–637, September 1976.
16. L.L. Horowitz and K.D. Senne, "Performance Advantage for Complex LMS for Controlling Narrowband Adaptive Arrays," *IEEE Trans. on Acous., Speech, and Signal Processing*, vol. ASSP-29, pp. 722–736, June 1981.

17. B. Fisher and N.J. Bershad, "The Complex LMS Adaptive Algorithm—Transient Weight Mean and Covariance with Applications to the ALE," *IEEE Trans. on Acous., Speech and Signal Processing*, vol. ASSP-31, pp. 34–45, February 1983.
18. S.T. Alexander and S.A. Rajala, "Optimal Gain Derivation for the LMS Algorithm Using a Visual Fidelity Criteria," *IEEE Trans. on Acous. Speech, and Signal Processing*, vol. ASSP-32, pp. 432–434, April 1984.
19. A.J. Berni, "Weight Jitter Phenomena in Adaptive Control Loops," *IEEE Trans. Aerosp. and Elec. Sys.*, vol. AES-14, pp. 355–361, July 1977.
20. N.J. Bershad and L.Z. Qu, "On the Joint Characteristic Function of the Complex Scalar LMS Weight," *IEEE Trans. on Acous., Speech and Signal Processing*, vol. ASSP-32, pp. 1166–1175, December 1984.
21. R.A. Iltis and L.B. Milstein, "An Approximate Statistical Analysis of the Widrow LMS Algorithm with Application to Narrow-Band Interference Rejection," *IEEE Trans. on Communications*, vol. COM-33, pp. 121–130, February 1985.
22. B. Widrow and S.D. Stearns, *Adaptive Signal Processing*, Prentice-Hall, Englewood Cliffs, NJ, 1985.
23. C.F.N. Cowan and P.M. Grant, *Adaptive Filters*, Prentice-Hall, Englewood Cliffs, NJ, 1985.
24. M.L. Honig and D.G. Messerschmitt, *Adaptive Filters*, Kluwer Academic Publishers, Hingham, MA, 1984.
25. S. Haykin, *Introduction to Adaptive Filters*, Macmillan, New York, 1984.
26. H.J. Trussell and J.D. Wang, "Cancellation of Harmonic Noise in Distribution Line Communications," *IEEE Trans. on Power Apparatus and Systems*, vol. PAS-104, pp. 3338–3344, December 1985.

CHAPTER 6
Applications of the LMS Algorithm

6.1 Introduction

The two general applications of system identification and linear prediction have been previously examined in parallel with analytical results in order to provide a physical basis for adaptive filtering. This chapter explores some additional applications to display the flexibility and versatility of adaptive methods. The applications of this chapter are by no means exhaustive, but are meant to provide more exposure to the use of adaptive signal processing. Additionally, the applications in this chapter are fairly straightforward and do not attempt to address all the specific implementation questions that are highly dependent on the specific hardware or software system used. More detailed system information can be gathered from the references listed.

All of the adaptive filters of this chapter are implemented with the least mean squares (LMS) algorithm. Several telecommunications applications have very successfully implemented the LMS algorithm in either chip set or very large scale integration (VLSI) [1–3], and its usage is quite mature in some areas. Section 6.2 investigates the application of adaptive filtering to the problem of echo cancellation on long distance telephone networks. Another example of adaptive filtering comes from waveform coding for communications. At the transmitter, LMS may be used as a linear predictor and a quantized version of the prediction error may be transmitted to the receiver. At the receiver, a high-fidelity reconstruction may be made using only the received prediction error without transmitting the predictor coefficients. This has been successfully used in speech and image coding [4–6], and is described in Section 6.3. The final example of adaptive signal processing examined in this chapter is in the area of spectrum estimation and modeling [7–9]. For

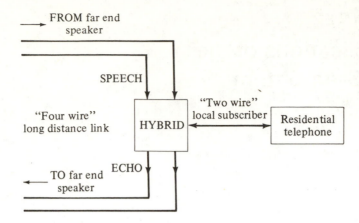

Figure 6.1 "Four-wire" to "two-wire" connection through the hybrid for the long distance telephone network.

signals that can be accurately modeled as autoregressive (AR) processes, the LMS algorithm provides a simple method of computing an AR spectral estimate. Section 6.4 describes how LMS can be used in these applications.

There are numerous other applications, such as adaptive beamforming [10], noise cancelling [11], and speech formant estimation [12]. For additional applications, the excellent introductory text by Widrow and Stearns [13] presents several different areas. In the area of antennas and beamforming, the text by Monzingo and Miller [13] focuses in detail on applications of adaptive filtering to array processing. Additionally, the text by Cowan and Grant [14] contains several applications of adaptive filtering to practical telecommunications problems.

6.2 Echo Cancellation

One of the main problems associated with long distance telephone communications is the generation of echoes due to impedance mismatches at various points in the telecommunications network [1–3]. The main source of echo is the junction between the "two-wire" local subscriber loop and the "four-wire" long distance link as illustrated in Figure 6.1. The links are not necessarily wire, but can be many types of media, such as fiber optic cable or the free space of a satellite link. The two-wire and four-wire links are joined at the central office by a device called the *hybrid*, which attempts to couple the energy from the far-end speaker to the subscriber, such that no speech energy is reflected back to the far end.

However, due to aging, component variance, unknown loop length, and other uncertainties, these impedances can rarely ever be balanced completely. For this reason, adaptive prediction filters are often used in the configuration

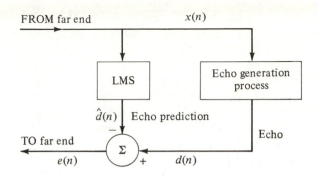

Figure 6.2   The LMS adaptive predictor used in the systems identification mode to cancel telephone echos.

shown in Figure 6.2 to cancel the unwanted echos. This is the systems identification mode previously considered.

A simple model for telephony echo generation is just a delay and attenuation [3], although more complicated models have been proposed [12]. Consider Figure 6.2, in which a speech waveform arrives as $x(n)$ and leaks through the hybrid to the return path as $d(n)$. A common assumption is that echo generation can be accurately modeled as an Nth order finite impulse response process, h_k. Thus, the echo output $d(n)$ of the system with input $x(n)$ is

$$d(n) = \sum_{k=0}^{N-1} h_k x(n-k). \qquad (6.2.1)$$

If the echo process can indeed be accurately modeled by (6.2.1), then the LMS algorithm can be used as in Figure 6.2 to make an accurate estimate of the echo output by

$$\hat{d}(n) = \sum_{k=0}^{N-1} w_k(n) x(n-k). \qquad (6.2.2)$$

The error then is

$$e(n) = d(n) - \hat{d}(n), \qquad (6.2.3)$$

from which (6.2.1) and (6.2.2) produce

$$e(n) = \sum_{k=0}^{N-1} [h_k - w_k(n)] x(n-k). \qquad (6.2.4)$$

The rationale for using LMS is now clear from (6.2.4); that is, minimizing the mean square error (MSE) should force the $w_k(n)$ to the values h_k. This will usually occur if the filter length N is large enough so that the length of the true echo process impulse response is less than N. Another consideration for actual systems is the initial arrival time of the echo energy. One method is to append additional coefficients to the adaptive filter, in which case it is easy to show that the $w_k(n) \approx 0$ for filter coefficients "earlier" than the echo arrival time. The first appreciably non-zero coefficient will then correspond to the echo arrival. For further information on other practical systems considerations, the reader is referred to [3, 14].

In echo cancellation, the main object is to drive $e(n)$ as small as possible during the start-up period, since during this phase, the return signal is almost completely echo before the local subscriber speaks. Figure 6.3 shows a section of a speech-like waveform and echo returned by a fifth-order finite impulse response echo model. This echo would be very noticeable if it returned on the connected path back to the speaker. Since the ear can distinguish between two sounds within about 50 ms of one another, it is necessary to cancel the echo by approximately 35 db in around 400 samples (50 ms at an 8 Khz sampling rate). Figure 6.3(c) shows the results of cancellation with a fifth-order LMS filter with $\alpha = 0.1$ (assuming known delay in the local subscriber loop). The echo in this case is cancelled by approximately 55 db with respect to $x(n)$. The initial echo energy will be passed, but would not be interpreted as speech by the far-end speaker. After convergence of the LMS filter, the echoes would be almost completely removed.

6.3 Adaptive Waveform Coding

Adaptive linear prediction filters have also found a great deal of application in the efficient encoding of speech and image waveforms for digital transmission over the local and long distance telephone networks. In fact, the International Telegraph and Telephone Consultative Committee (CCITT) has incorporated an adaptive predictor into the 32 Kbps standard for speech encoding to be used in telephone network applications [17, 18]. As an example of this approach to speech coding, this section examines the use of the LMS algorithm as the adaptive predictor in adaptive differential pulse code modulation (ADPCM). For more specific information on predictive encoding for speech systems, the interested reader is referred to the excellent papers by Gibson [4, 5] and the work by Honig and Messerschmitt [19], which evaluates several methods of adaptive speech encoding. Additionally, the LMS algorithm has also been used for adaptive image coding by Alexander and Rajala [6]. Whereas, in general, the channel quantizers in these type systems could also be made adaptive, this chapter focuses on the simpler case in which only the linear predictor is made adaptive. For more information on other methods of speech and image coding, the text by Jayant and Noll [20] contains a wide variety of techniques.

Figure 6.4(a) illustrates how an adaptive predictor, in general, and LMS, in particular, may be used in speech coding. A prediction, $\hat{x}(n)$, is made of the original speech, $x(n)$, from which the prediction error signal, $e(n)$, is generated:

$$e(n) = x(n) - \hat{x}(n) = x(n) - \sum_{i=1}^{N} w_i(n)\tilde{x}(n - i), \qquad (6.3.1)$$

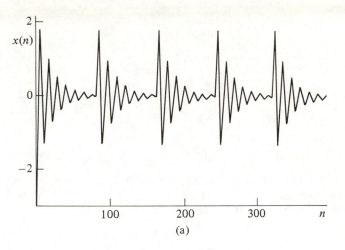

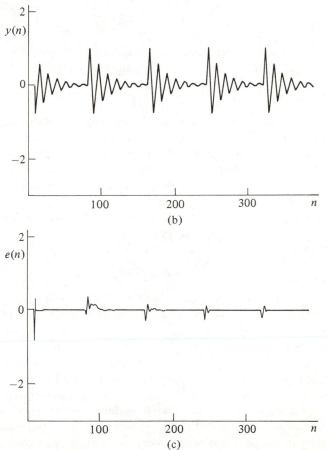

Figure 6.3 (a) Speech-like waveform input to the hybrid in echo cancellation example. (b) Echo output for fifth-order echo generation process. (c) Cancelled echo residual using LMS filter; $\alpha = 0.1$.

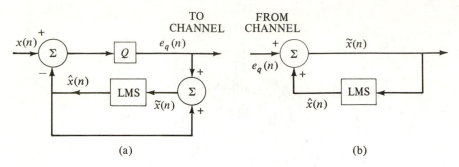

(a)   (b)

Figure 6.4   (a) The LMS algorithm used in the ADPCM encoder; quantized residual $e_q(n)$ sent to receiver.   (b) The LMS algorithm used in the ADPCM decoder at the receiver; errorless reception of $e_q(n)$ assumed.

where

$$\tilde{x}(n) = \hat{x}(n) + e_q(n) \tag{6.3.2}$$

is the reconstructed approximation to the original speech, $x(n)$. The prediction error is then quantized by the quantizer element Q, such that

$$e_q(n) = Q[e(n)] = e(n) + \eta_e(n). \tag{6.3.3}$$

In (6.3.2), the notation $Q[\cdot]$ represents the quantizer operation and $\eta_e(n)$ is the quantizer error. The complementary reconstruction filter or decoder shown in Figure 6.4(b) is located at the receiver and computes the reconstructed signal $\tilde{x}(n)$ based upon the (assumed errorless) reception of the $e_q(n)$.

In order to guarantee stability of this system, the reconstructed signal $\tilde{x}(n)$ must be used as the input to the predictors at both the transmitter and the receiver. It is easy to show that if both encoder and decoder predictors have the same initial conditions, then the only error in the reconstructed signal at the receiver is equal to the quantizer error, $\eta_e(n)$, or

$$\eta_e(n) = x(n) - \tilde{x}(n). \tag{6.3.4}$$

This is assuming errorless reception of the $e_q(n)$, which will be the case in the current application. Note that the result in (6.3.4) is independent of the specific algorithm used in the predictor. However, the cost of using an inefficient predictor is that the dynamic range of the resulting $e(n)$ might be quite large. This, in turn, would require a relatively large number of bits in the encoded version $e_q(n)$ to span the dynamic range of $e(n)$, such that $\eta_e(n)$ could be kept small. Since one of the design objectives of modern digital communications is to minimize the transmitted bit rate while maintaining acceptable distortion levels, the adaptive predictor is therefore a very beneficial approach.

Adaptivity in the predictor is necessary, since most practical information signals such as speech or images are highly nonstationary. Thus, the predictor may adapt its coefficients to "match" the changing statistical nature of the data. This will decrease the dynamic range of the error residual compared to the range of the residual produced by a fixed coefficient predictor. Encoding this residual will therefore require fewer quantizer levels to span the residual range for the adaptive predictor, which in turn implies fewer quantizer bits per residual sample to be sent over the channel. This is the basic approach taken by adaptive predictive coders to achieve a reduction in the amount of data sent over a band-limited channel.

As a specific example using the LMS adaptive predictor, consider the waveform $x(n)$ in Figure 6.5(a) that is to be encoded and transmitted to a receiver. As can be seen in the figure, the properties of the waveform change substantially, and a fixed coefficient prediction filter would require a channel quantizer with an unacceptably large number of bits per sample. Figure 6.5(b) displays the results of using the LMS adaptive predictor and a one bit per sample quantizer optimized for a Laplacian residual density function. This is a common method of quantizing the residual, since several investigations [20, 21] have shown that the Laplacian density accurately models the amplitude probability density function of the prediction residual. For a unity variance Laplacian-distributed signal, the optimum quantizer decision levels and output levels have been previously calculated [20, 21] for a number of different cases. For purposes of the present example, determination of the required quantizer parameters will therefore require only an estimate of the RMS value of the residual, since the unity variance parameters may then be scaled by the actual RMS value. There is definitely some distortion in the reconstructed waveform $\tilde{x}(n)$ at the receiver, but this is the cost of the data compression achieved. However, the reconstructed waveform resulting from using the fixed coefficient DPCM predictor and the same one bit Laplacian quantizer is shown in Figure 6.5(c). The DPCM predictor coefficients were computed using the statistics of the first segment of $x(n)$; as a result, the first segment of $\tilde{x}(n)$ in Figure 6.5(c) is constructed fairly well. However, the fixed predictor coefficients are extremely inaccurate for predicting the second segment of $x(n)$, and the reconstruction is extremely poor. Although a simple example, this type of performance is indicative of actual adaptive waveform coding systems. The reader is referred to [6] for examples of the performance gain to be expected for image coding using the LMS predictor.

6.4 Adaptive Spectrum Analysis

The adaptive linear predictor is also very useful in estimating the power spectrum of a signal or in tracking the dominant frequencies of a signal. This is applicable in many diverse areas, such as speech analysis and recognition [12], high-resolution spectral analysis [8, 15], and target tracking [22, 23]. The linear prediction filters examined thus far have all been finite impulse response (FIR) filters, or, equivalently, transversal all-zero filters. If the signal generation process can be accurately modeled as an all-pole spectral process,

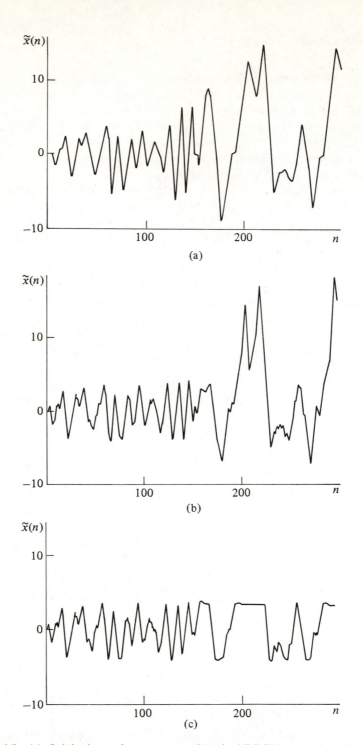

Figure 6.5 (a) Original waveform at transmitter in ADPCM example. (b) Reconstructed waveform at receiver using LMS predictor; $\alpha = 0.02$; 1 bit quantized residual transmitted. (c) Reconstructed waveform at receiver using fixed coefficient DPCM predictor; 1 bit quantized residual transmitted.

then FIR adaptive filters provide an excellent method of obtaining high-resolution spectral estimates and accurate frequency estimates for the dominant sinusoids.

To see this, assume that an information signal may be accurately modeled by the following difference equation:

$$x(n) = \sum_{k=1}^{N} a_k x(n-k) + v(n), \qquad (6.4.1)$$

where the a_k are called the *autoregressive* (*AR*) coefficients of the signal model, and $v(n)$ is an uncorrelated random process of power σ_v^2, which excites or drives the difference equation modeling the signal. The term "autoregressive" stems from the statistical literature [8], which in signal processing applications has become synonymous with an *all-pole* spectrum. A block diagram of such a signal generation process was shown in Figure 1.2. Note that the excitation $v(n)$ can be written as the nonpredictable portion of the signal or the *innovations* of the AR process:

$$v(n) = x(n) - \sum_{k=1}^{N} a_k x(n-k). \qquad (6.4.2)$$

It is this interpretation of $v(n)$ as the nonpredictable portion of $x(n)$ that suggests using the adaptive linear predictor for spectrum estimation.

It is easy to show that the frequency domain (ω-domain) representation of the power spectrum corresponding to the stochastic process given in (6.4.1) is given by [8, 16]:

$$S_{xx}(e^{j\omega}) = \frac{\sigma_v^2}{|A(e^{j\omega})|^2}, \qquad (6.4.3)$$

where

$$A(e^{j\omega}) = 1 - \sum_{k=1}^{N} a_k e^{-j\omega k}. \qquad (6.4.4)$$

A spectrum estimate of the type in (6.4.3) is not limited by the resolution constraints typical of Fourier-based spectrum estimates. Note that the estimate (6.4.3) can have extremely high magnitude, extremely narrow frequency peaks if the roots of the $A(e^{j\omega})$ polynomial are close to the unit circle. For this reason, these all-pole estimators are frequently referred to high-resolution spectrum estimates. However, this spectrum estimator is very dependent on the accuracy of the underlying method for estimating the AR coefficients a_k. If the estimated a_k are erroneous, then the roots of the resulting polynomial $A(e^{j\omega})$ may be substantially in error, resulting in a significantly degraded spectrum estimate. An excellent tutorial on high-resolution spectrum estimation that details benefits and exposes pitfalls is the paper by Kay and Marple [8].

In the current application to adaptive spectrum estimation, the data signal $x(n)$ is acquired from a sensor, and it is desired to compute an AR power spectrum of the underlying signal process from the acquired $x(n)$. The LMS adaptive filter used in the linear prediction mode provides a simple method of computing this spectrum. If $x(n)$ is predicted using the N most recent samples $x(n-1), \ldots, x(n-N)$, then the error $e(n)$ is given by

$$e(n) = x(n) - \sum_{k=1}^{N} w_k(n) x(n-k), \qquad (6.4.5)$$

where the $w_k(n)$ are the LMS coefficients. Note the similarity between (6.4.5) and (6.4.2). If the $w_k(n)$ were equal to the unknown a_k, then $e(n)$ would indeed be the true innovations $v(n)$. Thus, to accurately estimate the a_k, the LMS algorithm may be used to minimize $E\{e^2(n)\}$, which should force the $w_k(n)$ to the a_k. Corresponding to (6.4.4), the transform of the LMS based error filter is

$$A_n(e^{j\omega}) = 1 - \sum_{k=1}^{N} w_k(n) e^{-j\omega k}, \qquad (6.4.6)$$

where the "n" subscript on $A_n(e^{j\omega})$ signifies the computation is made using the LMS weights $w_k(n)$. At convergence, the $w_k(n)$ will approach the true a_k and $A_n(e^{j\omega})$ will approach the true $A(e^{j\omega})$.

However, recall from Chapter 5 that the $w_k(n)$ produced by the LMS algorithm are stochastic processes, and at convergence it is only the expectation $E\{w_k(n)\}$ that approaches the true a_k. The individual instantaneous $w_k(n)$ have a variance or jitter that is inversely porportional to the feedback gain parameter α. This variance will perturb the $w_k(n)$ from the actual a_k, with the result that the spectral estimate using (6.4.6) will fluctuate significantly as each iteration of LMS is performed. Therefore, some method is needed to "smooth" the fluctuations in the LMS coefficients such that an approximation to their expected values may be computed.

One simple method of doing this at LMS convergence is to find a short time average of the LMS weights, usually by some form of recursively updated average over the last K samples. For example, an approximation to the expected value $E\{w_k(n_0)\}$ may be made by finding the average of the K most recent values $w_k(n_0)$ through $w_k(n_0 - K + 1)$. In the example to follow, this simple method will be seen to greatly stabilize the resulting spectrum estimate using the LMS algorithm.

Hence, a simple method of computing the AR spectrum estimate is as follows:

(1) Use the adaptive filter (in this case, LMS) in the linear prediction mode.
(2) At convergence, compute a short-term time average of the $w_k(n)$, denoted as $\hat{w}_k$.
(3) Evaluate the function

$$A_n(e^{j\omega}) = 1 - \sum_{k=1}^{N} \hat{w}_k e^{-j\omega k}$$

for the desired frequencies over the range $0 \leq \omega \leq \pi$.

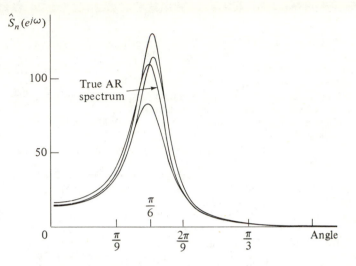

Figure 6.6   AR spectrum estimates computed using the LMS algorithm.

(4) Compute the power spectrum estimate

$$\hat{S}_n(e^{j\omega}) = \frac{1}{|A_n(e^{j\omega})|^2} \qquad (6.4.7)$$

Note that if N were a power of two, then the fast Fourier transform (FFT) could be used to compute the transform of the $w_k(n)$ in (6.4.6) at selected discrete frequencies $\omega_m = 2\pi m/N$. This would enhance the computational speed of the resulting spectrum estimate.

As one example of adaptive spectrum analysis, consider the problem of estimating the spectrum of a stationary second-order process that has poles at $(z_1, z_2) = 0.90e^{\pm j\pi/6}$. This signal is created by the difference equation

$$x(n) = 1.558x(n-1) - 0.81x(n-2) + \sigma_v v(n), \qquad (6.4.8)$$

and has the power spectrum

$$S_{xx}(e^{j\omega}) = \frac{\sigma_v^2}{|1 - 1.558e^{-j\omega} + 0.81e^{-j2\omega}|^2}. \qquad (6.4.9)$$

where σ_v^2 is the power of the uncorrelated excitation signal $v(n)$.

Figure 6.6 shows the true AR spectrum $S_{xx}(e^{j\omega})$ and the estimates computed using a two- coefficient LMS filter for several different realizations of the stochastic signal $x(n)$. Additionally, an average of 20 $w_k(n)$ during the converged period were used to approximate the expectation of $w_k(n)$. As can be seen, the LMS-based spectrum is quite accurate in the center frequency estimate, although the magnitude of the spectrum is somewhat variable.

REFERENCES

1. C.W.K. Gritton and D.W. Lin, "Echo Cancellation Algorithms," *IEEE ASSP Magazine*, vol. 1, no. 2, pp. 30–38, April 1984.
2. M.M. Sondhi and D.A. Berkley, "Silencing Echoes on the Telephone Network," *Proceedings of the IEEE*, vol. 68, pp. 948–963, August 1980.
3. D.L. Duttweiler and Y.S. Chen, "A Single Chip VLSI Echo Canceller," *Bell Sys. Tech. J.*, vol. 59, no. 2, pp. 149–160, February 1980.
4. J.D. Gibson, "Adaptive Prediction in Speech Differential Encoding Systems," *Proceedings of the IEEE*, vol. 68, pp. 488–525, April 1980.
5. J.D. Gibson, "Adaptive Prediction for Speech Encoding," *IEEE ASSP Magazine*, vol. 1, no. 3, pp. 12–26, July 1984.
6. S.T. Alexander and S.A. Rajala, "Image Compression Results Using the LMS Adaptive Algorithm," *IEEE Trans. on Acous., Speech and Signal Processing*, vol. ASSP-33, pp. 712–715, June 1985.
7. L.J. Griffiths, "Rapid Measurement of Instantaneous Frequency," *IEEE Trans. on Acous., Speech and Signal Processing*, vol. ASSP-23, pp. 209–222, April 1975.
8. S.M. Kay and S.L. Marple, "Spectrum Analysis—A Modern Perspective," *Proceedings of the IEEE*, vol. 69, pp. 1380–1419, November 1981.
9. W.S. Hodgkiss and J.A. Presley, "Adaptive Tracking of Multiple Sinusoids Whose Power Levels are Widely Separated," *IEEE Trans. on Acous., Speech and Signal Processing*, vol. ASSP-29, pp. 710–721, June 1981.
10. B. Widrow, et al., "Adaptive Antenna Systems," *Proceedings of the IEEE*, vol. 55, pp. 2143–2159, December 1967.
11. B. Widrow, et al., "Adaptive Noise Cancelling: Principles and Applications," *Proceedings of the IEEE*, vol. 62, pp. 1692–1716, December 1975.
12. J. D. Markel, "Digital Inverse Filtering—A New Tool for Formant Trajectory Estimation," *IEEE Trans. Audio Electroacoust.*, vol. AU-20, pp. 129–137, June 1972.
13. R.A. Monzingo and T.W. Miller, *Introduction to Adaptive Arrays*, John Wiley & Sons, New York, 1980.
14. C.F.N. Cowan and P.M. Grant, *Adaptive Filters*, Prentice-Hall, Englewood Cliffs, NJ, 1985.
15. D.G. Childers, *Modern Spectrum Analysis*, IEEE Press, New York, 1978.
16. J. Makhoul, "Linear Prediction: A Tutorial Review," *Proceedings of the IEEE*, vol. 63, pp. 561–580, April 1975.
17. CCITT COM XVIII-R28, Working Party XVIII/2 Report (Speech Processing), Geneva, Switzerland, November 21–25, 1983.
18. W.R. Daumer, P. Mermelstein, X. Maitre, and I. Tokizawa, "Overview of the ADPCM Coding Algorithm," 1984 IEEE Global Telecommunications Conf., Atlanta, GA, November 1984.
19. M.L. Honig and D.G. Messerschmitt, "Comparison of Adaptive Linear Prediction Algorithms in ADPCM," *IEEE Trans. on Communications*, vol. COM-30, pp. 1775–1785, July 1982.
20. M.D. Paez and T.H. Glisson, "Minimum Mean Squared Error Quantization in Speech PCM and DPCM Systems," *IEEE Trans. on Communications*, vol. COM-20, pp. 225–230, April 1972.
21. N.S. Jayant and P. Noll, *Digital Coding of Waveforms*, Prentice-Hall, Englewood Cliffs, NJ, 1984.
22. L.J. Griffiths, "Rapid Measurement of Instantaneous Frequency," *IEEE Trans. on Acous., Speech and Signal Processing*, vol. ASSP-23, pp. 209–222, April 1975.
23. N.J. Bershad, P.L. Feintuch, F.A. Reed, and B. Fisher, "Tracking Characteristics of the LMS Adaptive Line Enhancer Response to a Linear Chirp Signal in Noise," *IEEE Trans. Acous., Speech and Signal Processing*, vol. ASSP-28, pp. 504–516, October 1980.

CHAPTER 7
Gradient Adaptive Lattice Methods

7.1 Introduction

The transversal filter structure was seen in Chapter 2 to be one form for implementing the desired Nth order linear prediction filter. Then in Chapter 5, the LMS adaptive filter algorithm was seen to be a natural candidate for implementing the transversal form for an adaptive linear prediction filter. However, in Chapter 3, an alternative structure denoted as the lattice formulation was seen to be a direct result of using Durbin's method for solving the normal equations. To use the lattice filter for processing actual data required that the reflection coefficients, k_p, $1 \leq p \leq N$, be computed based upon estimates of the autocorrelation coefficients. These autocorrelation estimates were used in the very efficient Durbin's algorithm for producing the k_p. For additional review on the autocorrelation method of linear prediction and other fixed coefficient approaches, the reader is referred to the excellent article by Makhoul [1].

However, this approach is somewhat of an indirect procedure, since it is first required to estimate the autocorrelation coefficients and then use Durbin's recursion. Moreover, if there is error in estimating the autocorrelation coefficients, then this error will propagate directly in erroneous reflection coefficient computations. It would be more efficient if these reflection coefficients could be estimated directly from the data and possibly updated on a sample-by-sample basis, such as was done for the least mean squares (LMS) algorithm. There are several ways of accomplishing this, incorporating several levels of computational complexity, and this is the main topic of this chapter. Several of the techniques in this chapter are denoted as adaptive lattice methods. However, these methods should not be confused with the very powerful family of least squares lattice (LSL) algorithms, which are also adaptive. The LSL algorithms are the subject of Chapter 10.

Many developments in the gradient adaptive lattice occurred as the result of the sensitivity of LMS to the eigenvalue disparity or ill conditioning of many signal processing environments. Griffiths [12] originally applied the adaptive lattice to noise cancelling as an alternative to transversal adaptive filtering. Carter [13] then investigated using the gradient adaptive lattice in speech coding. Another application was by Satorius and Alexander [9], who applied the adaptive lattice to achieve a more rapid convergence for equalizers operating in highly dispersive channels. In another area, Gibson and Haykin [14] have done considerable work on applying the lattice filter for clutter rejection and discrimination in radar signal processing, and an excellent tutorial on these uses has been done recently by Haykin [15]. Concerning the performance analysis of the gradient adaptive lattice, Medaugh and Griffiths [16], Honig and Messerschmitt [8], and Gibson and Haykin [17] have investigated the convergence properties of the gradient lattice. Although there are some special cases in which the gradient transversal and lattice filters give similar convergence speeds, for the majority of practical applications, the lattice offers a more rapid convergence than does the transversal LMS algorithm. Similarly, the convergence of the gradient lattice, in general, is not as rapid as the exact least squares methods to be examined in Chapters 10 and 11. The gradient lattice therefore represents an approach that is usually more rapidly converging than LMS, but not as computationally intensive as the least squares methods.

This chapter is organized as follows. Section 7.2 discusses some methods for computing the reflection coefficients directly from the data, as well as the performance measures for these methods. The methods discussed in Section 7.2 are all fixed coefficient methods; that is, once the k_p are computed, they are held fixed for the duration of processing. However, they are computed using the acquired data and not by Durbin's algorithm. The transition to time-recursive adaptive computation of the reflection coefficients is the topic of Section 7.3. A very popular method known as the gradient adaptive lattice is derived and discussed. This method has faster convergence properties than the LMS algorithm, with only a moderate increase in computational complexity. Finally, some performance comparisons between gradient adaptive lattice methods and LMS are discussed in Section 7.4.

7.2 Lattice Reflection Coefficient Computation

This chapter investigates methods of adaptively updating the lattice form of the linear prediction filter. A Nth order linear prediction lattice filter consists of N sequential stages, as shown in Figure 3.1(a). The operation of the pth stage of the lattice $(1 \leq p \leq N)$ is shown in Figure 3.1(b) and is defined by the input-output relations (3.3.23) and (3.3.24), repeated here,

$$e_p^f(n) = e_{p-1}^f(n) - k_p e_{p-1}^b(n-1), \qquad (3.3.23)$$

$$e_p^b(n) = e_{p-1}^b(n-1) - k_p e_{p-1}^f(n), \qquad (3.3.24)$$

where $e_p^f(n)$ and $e_p^b(n)$ are the pth order forward predictor error (FPE) and backward predictor error (BPE), respectively, at time n. In previous work with the lattice structure in Chapter 3, the autocorrelation coefficients and equations (3.2.18) were used to compute the reflection coefficients k_p when Durbin's algorithm was used to solve the normal equations. In the present chapter, however, the interest is in computing the k_p directly from the data. One requirement of the candidate computation is that adding the "new" pth stage to the "old" $p-1$ stage lattice must implement the minimum mean square error (MSE) linear predictor of order p.

Recall from Table 3.1 in solving the normal equations via Durbin's algorithm that at the end of the pth iteration, the pth order minimum MSE filter $\mathbf{w}_p$ could be derived. One property of $\mathbf{w}_p$ is that it provides the lowest power prediction error sequence of any linear filter. Therefore, when the minimum MSE linear prediction filter is implemented on real data using the lattice form, then the FPE and BPE sequences in Figure 3.1 at the output of each stage must have the minimum power that they can obtain. Hence, when considering the pth order linear prediction filter, the $\mathbf{w}_p$ should be chosen to minimize some function of the pth order FPE and/or BPE powers. But from Table 3.1 listing Durbin's algorithm, it is seen that the $\mathbf{w}_p$ may be calculated from a knowledge of the reflection coefficients, k_p. Therefore, for the Nth order linear prediction lattice filter, the k_p should also be chosen to minimize some function of the FPE and/or BPE power at each stage of the lattice.

Thus, another minimization problem is suggested; namely, that of minimizing a prediction error power output from the pth stage. As has now been done several times in this text, the procedure is a standard one in minimization and involves calculating a gradient, setting it to zero, and then solving for the required values of k_p. However, since in the lattice there are now two error sequences, $e_p^f(n)$ and $e_p^b(n)$, at each stage the question immediately arises of which error sequence to use in the minimization. There are several error measures that can be constructed from these sequences. For example, one possibility would be to find the set of k_p that minimize the mean square power of the FPE by solving

$$\frac{\partial}{\partial k_p} E\{[e_p^f(n)]^2\} = 0. \qquad (7.2.1)$$

On the other hand, another possibility is to find the k_p that minimize the mean square power of the BPE sequence by solving

$$\frac{\partial}{\partial k_p} E\{[e_p^b(n)]^2\} = 0. \qquad (7.2.2)$$

Indeed, error measures such as these are often used, and many of them have been investigated by Makhoul [1, 4]. However, one method in particular results in a symmetric form for the resulting k_p. This approach will be examined in the remainder of this section.

Define the squared error sum $\sigma_p(n)$ as

$$\sigma_p(n) = [e_p^f(n)]^2 + [e_p^b(n)]^2. \qquad (7.2.3)$$

Minimizing the expectation of $\sigma_p(n)$ with respect to the unknown k_p then results in solving

$$\frac{\partial}{\partial k_p} E\{\sigma_p(n)\} = 0. \qquad (7.2.4)$$

The error measure in (7.2.3) was first used by Burg [2, 3] in geophysical applications of linear prediction filtering. This minimization is straightforward, although somewhat lengthy, and is examined in the problems at the end of the chapter. The resulting expression for the reflection coefficients is given by

$$k_p = \frac{2E\{e_{p-1}^f(n)e_{p-1}^b(n-1)\}}{E\{[e_{p-1}^f(n)]^2\} + E\{[e_{p-1}^b(n-1)]^2\}}. \qquad (7.2.5)$$

While the expression (7.2.5) is very important from a theoretical standpoint, it is limited in its computational usefulness. Note that (7.2.5) still requires using the statistical concept of expectation in order to compute the optimal reflection coefficients k_p. This is similar to the situation in Chapters 2 and 3 in which expectations were required to solve the normal equations for the optimal predictor $\mathbf{w}_N^*$. However, in signal processing applications using real data, there is rarely an exact knowledge of the true statistics of the underlying random processes.

One method previously examined in Section 3.2 employed block-processing methods for estimating the true autocorrelation coefficients of a stationary, ergodic process. A similar approach could now be taken in order to estimate the k_p in (7.2.5), but the stage-by-stage construction of the lattice presents another problem. To see this, note that (7.2.5) gives the reflection coefficients at the pth stage in terms of the error signals out of the previous $p - 1$st stage. Therefore, the sequences $e_{p-1}^f(n)$ and $e_{p-1}^b(n)$ could be acquired over an entire block, and then estimates of the expectations required in (7.2.5) could be made as follows:

$$\hat{E}\{[e_{p-1}^f(n)]^2\} = \frac{1}{K} \sum_{k=1}^{K} [e_{p-1}^f(k)]^2, \qquad (7.2.6)$$

$$\hat{E}\{[e_{p-1}^b(n-1)]^2\} = \frac{1}{K} \sum_{k=1}^{K} [e_{p-1}^b(k-1)]^2 \qquad (7.2.7)$$

$$\hat{E}\{e_{p-1}^f(n)e_{p-1}^b(n-1)\} = \frac{1}{K} \sum_{k=1}^{K-1} e_{p-1}^f(k)e_{p-1}^b(k-1), \qquad (7.2.8)$$

where there are a total of K samples in the block of data.

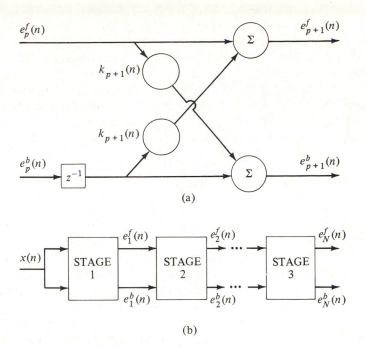

Figure 7.1 (a) Single stage of the gradient adaptive lattice filter. (b) Cascade of stages to form Nth order gradient adaptive lattice filter.

Although theoretically appropriate, there is one obvious drawback to this approach. One block of K samples must be acquired and averages computed for the first stage k_1 before the computations for k_2 may begin. Then, second stage errors must be computed (and another delay of K samples incurred) before the computation of k_3 may begin, and so on. This direct approach leads to a delay of NK sample times before the reflection coefficients of the entire N stage lattice may be computed. This delay may well be prohibitive for the changing signal environments encountered in most signal-processing applications.

Thus, this approach is not usually taken in practice due to the delay and computational cost of implementing the sum of products in each of (7.2.6)–(7.2.8). In many signal-processing applications, such as low bit rate speech transmission [4–7] and spectral analysis [1,3], the following procedure is often used. First, estimators of the type found in Chapters 2 and 3 are used to compute autocorrelation function estimates directly from the acquired data. These autocorrelation function estimates are next used in Durbin's algorithm (or other closely related fast algorithms) to solve for the relection coefficients k_p. Lastly, these k_p are then inserted in the lattice structure of Figure 7.1, which has $x(n)$ as input and the Nth order FPE and BPE as output. This set of k_p could then be used to process the signal as long as no major change in signal statistics were perceived. However, many real-time oriented applications, such as linear predictive coding (LPC) for speech, compute a new set of k_p for each new K sample block of input data.

This block-oriented approach has been used often in speech [1, 7], in which the short-term stationarity of speech requires a new set of reflection coefficients to be calculated at 10–30 ms intervals. However, not all signal environments will have such a known short time stationarity, and it is therefore necessary to investigate methods for updating the lattice parameters recursively; that is, as each new input data sample appears.

7.3 Adaptive Lattice Derivations

As seen in the previous section, there is often a need to update the reflection coefficients. If the input sequence $x(n)$ were truly stationary, then only one calculation of the reflection coefficients need be performed. However, if the input $x(n)$ has a nonstationary character and the blockwise autocorrelation method is not able to track the nonstationarity, then one alternative is to adaptively update the reflection coefficients with each new time sample n.

The strategy is similar to that employed in the LMS algorithm. A linear filter will be used to predict a desired signal, this predicted value will be compared to the actual value of the signal to generate an error, and the error will be used to update the filter coefficients. Once the lattice filter structure has been chosen, then the only unknowns are the reflection coefficients; therefore, the coefficients of the adaptive filters in this realization are the reflection coefficients. However, unlike the case of LMS, several methods have been developed for adaptively estimating the reflection coefficients. Some of these methods are discussed in this section.

Gradient lattice algorithm

One method of adaptively updating the reflection coefficients directly from the available error sequences uses the strategy of the LMS method of Chapter 5. Recall the update of (5.3.4),

$$\mathbf{w}_N(n + 1) = \mathbf{w}_N(n) - \mu \nabla_\mathbf{w}[e^2(n)], \tag{5.3.4}$$

where $e^2(n)$ is the instantaneous prediction error. Specifically, (5.3.4) states that the predictor solution at time $n + 1$ is obtained by updating the solution at time n in the direction opposite to the gradient of the squared prediction error. Following this approach, a gradient-based method for updating the reflection coefficients may be written as

$$k_p(n + 1) = k_p(n) - \mu \nabla_{k_p} \sigma_p(n), \tag{7.3.1}$$

where $\sigma_p(n)$ is the sum of squared BPEs and FPEs and is given in (7.2.3). The parameter μ is a gain constant, similar to that used in the update for LMS.

Therefore, the new reflection coefficient is obtained by updating the old coefficient in a direction opposite to the gradient of the total prediction error power at the pth stage.

Since $k_p(n)$ is a scalar, then each $k_p(n)$ may be optimized separately. Therefore,

$$\nabla_{k_p} \sigma_p(n) = \frac{\partial}{\partial k_p} \sigma_p(n), \tag{7.3.2}$$

where the notation has been simplified such that $\partial/\partial k_p = \partial/\partial k_p(n)$. Simple algebra then gives

$$\nabla_{k_p} \sigma_p(n) = 2\left\{ e_p^f(n) \frac{\partial}{\partial k_p} e_p^f(n) + e_p^b(n) \frac{\partial}{\partial k_p} e_p^b(n) \right\}. \tag{7.3.3}$$

Carrying out the derivative and using (7.2.1) then gives the following recursion:

$$k_p(n+1) = k_p(n) + \beta[e_p^f(n)e_{p-1}^b(n-1) + e_p^b(n)e_{p-1}^f(n)], \tag{7.3.4}$$

where $\beta = 2\mu$ is the gain constant. The adaptivity of the $k_p(n)$ now gives the block diagram shown in Figure 7.1 for a single stage of the lattice.

Note that (7.3.4) is similar to the LMS update from (5.3.4). Therefore, the relative size of the gain β must be considered such that the resulting algorithm is stable. Additionally, (7.3.4) is often implemented with β replaced by a time-varying gain $\beta(n)$, which will mitigate the problems of instability. To formulate this gain normalization, it is useful to first consider another method [4, 8] of recursively updating the reflection coefficients. In this method, the expectations in (7.2.5) are first approximated by a summation over the amount of data acquired at time n; that is, at time n

$$k_p(n) = \frac{2N_p(n)}{D_p(n)}, \tag{7.3.5}$$

where

$$N_p(n) = \sum_{k=1}^{n} e_{p-1}^f(k)e_{p-1}^b(k-1), \tag{7.3.6}$$

$$D_p(n) = \sum_{k=1}^{n} \{[e_{p-1}^f(k)]^2 + [e_{p-1}^b(k-1)]^2\}. \tag{7.3.7}$$

Using these definitions thus produces

$$N_p(n+1) = e_{p-1}^f(n+1)e_{p-1}^b(n) + N_p(n), \tag{7.3.8}$$

$$D_p(n+1) = [e_{p-1}^f(n+1)]^2 + [e_{p-1}^b(n)]^2 + D_p(n). \tag{7.3.9}$$

The update for $k_p(n+1)$ is therefore easily seen to be given by

$$k_p(n+1) = \frac{2N_p(n+1)}{D_p(n+1)} = \frac{2[N_p(n) + e_{p-1}^f(n+1)e_{p-1}^b(n)]}{D_p(n) + [e_{p-1}^f(n+1)]^2 + [e_{p-1}^b(n)]^2}. \tag{7.3.10}$$

Although (7.3.10) is valid in its present form, it can be rearranged into another form that explicitly shows $k_p(n)$. Dividing the numerator of (7.3.10) by its denominator produces after some simplification

$$k_p(n + 1) = \left[\frac{D_p(n)}{D_p(n + 1)} \right] k_p(n) + 2e^f_{p-1}(n + 1)e^b_{p-1}(n). \qquad (7.3.11)$$

Note that (7.3.11) is a stable algorithm as long as $D_p(n + 1) \neq 0$. For all practical cases, the condition will be met since, from (7.3.7), all the summed terms of $D_p(n + 1)$ are greater than zero if $e^f_{p-1}(n)$ and $e^b_{p-1}(n - 1)$ are not instantaneously zero. Furthermore, the computation in (7.3.10) shows that $D_p(n) \leq D_p(n + 1)$, with the equality holding only if $e^f_{p-1}(n + 1)$ and $e^b_{p-1}(n)$ are instantaneously zero. This essentially guarantees the convergence of (7.3.11), since it may be shown [8] that the algorithm (7.3.11) converges for $D_p(n) < D_p(n + 1)$.

Normalized gradient lattice algorithm

The main drawback associated with the gradient lattice algorithm (7.3.4) for updating $k_p(n)$ is that the feedback gain parameter β must be judiciously chosen to guarantee stability. However, a time-varying value of β may be chosen such that β is effectively normalized to the power of the signals and thus avoids the problems associated with filter instability. One such method applied by Satorius and Alexander [9] to channel equalization is to replace β by $1/d_p(n)$, where

$$d_p(n) = [1 - \alpha]d_p(n - 1) + [e^f_p(n)]^2 + [e^b_p(n)]^2, \qquad (7.3.12)$$

where $\alpha \ll 1$, in this case, is a small constant. Therefore, the gradient adaptive lattice algorithm becomes

$$k_p(n + 1) = k_p(n) + \frac{1}{d_p(n)} [e^f_p(n)e^b_{p-1}(n - 1) + e^b_p(n)e^f_{p-1}(n)], \quad (7.3.13)$$

with $d_p(n)$ recursively updated as in (7.3.12). This definition of $d_p(n)$ is similar to the "normalized gradient" or "normalized LMS" algorithm that has been used extensively [10, 11] in applications such as echo cancellation. From (7.3.12) it is seen that the value $d_p(n)$ decreases for small prediction error, which indicates an increase in the multiplying gain $1/d_p(n)$. This is quite desirable since low prediction error implies a high degree of accuracy in the predictor, or that the linear filter has indeed formed an adequate model of the system under consideration. Therefore, any increase in the prediction errors should be due to a change in system parameters, and it is desirable to have the adaptive lattice adjust to this new system very quickly. A large gain value $1/d_p(n)$ will aid in giving an initially rapid convergence to the new system parameters. Conversely, should there be a processing environment consisting of a high white noise level in addition to any signal, then the prediction errors

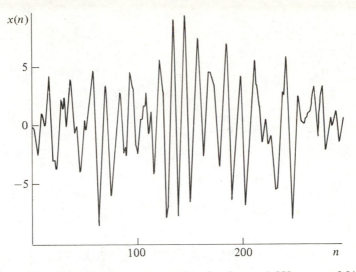

Figure 7.2 Second-order autoregressive signal; $a_1 = 1.558$, $a_2 = -0.81$.

could be quite large. This, in turn, would cause $d_p(n)$ to be quite large, and therefore the gain term $1/d_p(n)$ would be fairly small. Hence, the adaptive lattice would not respond very rapidly to the high noise environment.

7.4 Performance Example

This section presents a simple example that demonstrates the potential gain in convergence speed of the gradient adaptive lattice compared to the LMS algorithm. However, this speed-up in convergence is not without some cost, since there are added computational operations and storage locations that must be accommodated. This cost is further explored in the problems at the end of the chapter.

The example in this section is the now familiar one of determining the autoregressive (AR) coefficients that have generated a data signal. The signal $x(n)$ shown in Figure 7.2 was created by passing white uncorrelated gaussian noise $v(n)$ through the linear system described by

$$x(n) = 1.558x(n-1) - 0.81x(n-2) + v(n). \tag{7.4.1}$$

This is the AR model of (1.2.5) with $a_1 = 1.558$ and $a_2 = -0.81$. If the LMS algorithm is used in the linear prediction, then the first two LMS filter coefficients should converge in the mean to the coefficients in (7.4.1). That is, $E\{w_1(n)\} \to 1.558$ and $E\{w_2(n)\} \to -0.81$. Since the lattice predictor also implements the minimum MSE linear prediction filter, it is also possible to extract the AR coefficients from the reflection coefficients of the adaptive

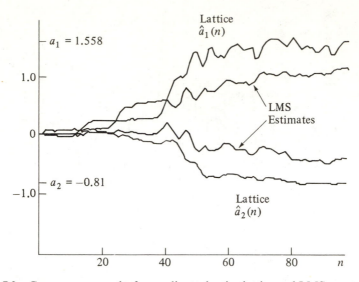

Figure 7.3   Convergence results for gradient adaptive lattice and LMS; second-order autoregressive signal; $a_1 = 1.558$, $a_2 = -0.81$.

lattice. It is easy to show that these AR estimates are given by

$$\hat{a}_1(n) = k_1(n)[1 - k_2(n)], \tag{7.4.2a}$$

$$\hat{a}_2(n) = k_2(n). \tag{7.4.2b}$$

The simple gradient adaptive lattice incorporating the reflection coefficient update of (7.3.4) was used to extract the estimated AR coefficients of $x(n)$. A gain parameter of $\beta = 0.01$ was used in (7.3.4) and produced the convergence curve shown in Figure 7.3. For comparison, the "best" LMS estimate is shown as well. This was the highest α level ($\alpha = 0.01$) achievable before instability occurred. As can be seen, the convergence is somewhat more rapid using the lattice than for LMS, while the misadjustment is approximately the same for both lattice and LMS. This is indicative of the performance characteristics achievable in many applications of the lattice versus LMS.

This chapter has introduced the concept of the adaptive lattice and appealed to the minimization of stagewise prediction error by gradient methods to derive the gradient adaptive lattice. However, in Chapter 10, it will be seen that a more powerful lattice method known as the least squares lattice (LSL) will provide even more rapid convergence. Furthermore, the implementation of LSL will be seen to require only a moderate increase in computations over the gradient lattice.

PROBLEMS

1. In this problem, assume the lattice filter structure of Figure 7.1, in which the selection of the reflection coefficients k_{p+1} determine the performance of the lattice.

(a) Derive an expression for the optimal reflection coefficients that results from minimizing the forward error measure $E\{[e_p^f(n)]^2\}$ at each stage of the lattice.

(b) Now derive an expression for the optimal reflection coefficients that minimizes the backward error measure $E\{[e_p^b(n)]^2\}$ at each lattice stage.

(c) Now find an expression for the optimal reflection coefficients that results from minimizing the sum squared error measure of (7.2.3). Verify that the result is given by (7.2.5).

2. In the recursive lattice algorithm of (7.3.10), verify that the update of (7.3.11) for $k_p(n + 1)$ is indeed given by the recursion (7.3.11).

3. Consider a running estimate of the sum squared prediction error power as in (7.3.12). Assume that the expected value $E\{[e_p^f(n)]^2 + [e_p^b(n)]^2\}$ has been a constant value, C, for a long time prior to $n = n_0$.

(a) What is the expected value of $d_p(n)$ in (7.3.12)?

(b) Assume that at $n = n_0 + 1$, the signal environment changes such that $E\{[e_p^f(n)]^2 + [e_p^b(n)]^2\} = 10\,C$. Find the trajectory of $E\{d_p(n)\}$ as it converges to its new value.

4. Compute the computational complexity and storage requirements for the three lattice algorithms with $k_p(n)$ updates given by (7.3.4), (7.3.11), and (7.3.13). Be sure to include the computations involved in implementing each lattice stage. Compare this with the computations and storage requirements for the Nth order transversal LMS filter.

REFERENCES

1. J. Makhoul, "Linear Prediction: A Tutorial," *Proceedings of the IEEE*, vol. 63, pp. 561–580, April 1975.
2. J.P. Burg, "Maximum Entropy Spectral Analysis," Ph.D. Dissertation, Geophysics Department, Stanford University, 1975.
3. D.G. Childers, ed., *Modern Spectrum Analysis*, IEEE Press, New York, 1978.
4. J. Makhoul, "Stable and Efficient Lattice Methods for Linear Prediction," *IEEE Trans. Acous., Speech, and Signal Processing*, vol. ASSP-25, no. 5, pp. 423–428, October 1977.
5. L.R. Rabiner and R.W. Schafter, *Digital Processing of Speech*, Prentice-Hall, Englewood Cliffs, NJ, 1978.
6. J.D. Markel and A.H. Gray, *Linear Prediction of Speech*, Springer-Verlag, New York, 1975.
7. S.T. Alexander, "A Simple Noniterative Speech Excitation Algorithm Using the LPC Residual," *IEEE Trans. on Acous., Speech, and Signal Processing*, vol. ASSP-33, pp. 432–434, April 1985.
8. M.J. Honig and D.D. Messerschmitt, "Convergence Properties of an Adaptive Digital Lattice Filter," *IEEE Trans. Acous., Speech, and Signal Processing*, vol. ASSP-29, pp. 642–653, June 1981.
9. E.H. Satorius and S.T. Alexander, "Channel Equalization Using Adaptive Lattice Algorithms," *IEEE Trans. on Communications*, vol. COM-27, pp. 899–905, June 1979.
10. D.L. Dutweiler, "A Twelve Channel Digital Echo Canceller," *IEEE Trans. on Communications*, vol. COM-26, pp. 647–653, May 1978.
11. M.M. Sondhi and D.A. Berkley, "Silencing Echoes on the Telephone Network," *Proceedings of the IEEE*, vol. 68, pp. 948–963, August 1980.

12. L.J. Griffiths, "A Continuously Adaptive Filter Implemented as a Lattice Structure," *Proceedings*, 1977 IEEE Int. Conf. on Acous., Speech, and Signal Processing, Hartford, CT, April 1977.
13. T.E. Carter, "Study of an Adaptive Lattice Structure for Linear Prediction Analysis of Speech," *Proceedings*, 1978 IEEE Int. Conf. on Acous., Speech, and Signal Processing, Tulsa, OK, May 1978.
14. C. Gibson and S. Haykin, "Radar Performance Studies of Adaptive Lattice Clutter Suppression Filters," *Proc. Inst. Elec. Eng., London*, vol. 130, part F, pp. 357–367, 1983.
15. S. Haykin, "Radar Signal Processing," *IEEE ASSP Magazine*, vol. 2, pp. 2–18, April 1985.
16. R.S. Medaugh and L.J. Griffiths, "A Comparison of Two Fast Linear Predictors," *Proceedings*, 1981 IEEE Int. Conf. on Acous., Speech, and Signal Processing, Atlanta GA, May 1981.
17. C. Gibson and S. Haykin, "Learning Characteristics of Adaptive Lattice Filtering Algorithms," *IEEE Trans. on Acous., Speech and Signal Processing*, vol. ASSP-28, pp. 681–691, December 1980.

CHAPTER 8
Recursive Least Squares Signal Processing

8.1 Introduction

The adaptive signal processing methods developed thus far in the text have all addressed the problem of solving the normal equations as given by (2.3.5). Implicit in this approach is that the *statistical* error measure of mean square prediction error is being minimized. The least mean squares (LMS) algorithm of Chapter 5 emulated the approach of the method of steepest descent for solving the normal equations, whereas the gradient lattice method of Chapter 7 used Durbin's algorithm as a basis for computing an order-recursive solution to (2.3.5).

However, most adaptive algorithms do not have access to the unknown autocorrelation functions, nor do they, in general, try to estimate the auto-correlation function. Recall that the adaptive signal processing methods examined thus far only use the actual data to compute predictions and update their coefficients. This leads to the main deficiency of the methods so far examined; namely, they have been derived using a statistical error measure, whereas their implementation utilizes the exact data available in the specific application. Therefore, it is quite logical to consider prediction filters based upon error measures derived from the exact data signal acquired.

This approach leads to the family of extremely powerful adaptive signal-processing techniques known collectively as *recursive least squares* (RLS) techniques. In the current chapter, the RLS filter using matrix inverse techniques will be examined. This method is sometimes known as the Kalman algorithm [1–3] by virtue of its similarity to the state space stochastic filter developed by Kalman [4]. In the remainder of the text, attention will be focused on those methods that most clearly convey the underlying mathematical principles and, at the same time, have also been readily incorporated into applications. The current chapter examines the general RLS filter, whose properties have been thoroughly investigated and found to be useful in channel equalization [1, 2], systems identification [5], speech analysis [15], and numerous other areas [6]. This method will be referred to simply as the RLS method in this chapter.

The general problem of finding the optimal least squares filter for a given set of acquired data has a long and rich history, and its origins are usually attributed to Gauss and the estimation of asteroid orbits [7]. There are two general categories of least squares methods: (1) block-processing techniques, and (2) recursive-processing techniques. In block-oriented techniques, a block or frame of K data samples is first acquired, and then computation of the least squares filter is done off-line. While extremely powerful for filtering, smoothing, and identification of processes in applications that do not require a real-time capability, block methods are not often candidates for real-time signal processing since there is an inherent delay in acquiring the necessary samples and then in computing the block of output data. The RLS techniques of current research interest implement the same least squares mathematics as block techniques, but instead perform the implementation in a sequential or recursive manner, thus avoiding many of the problems of delay and storage associated with the block techniques. Since the remainder of this text focuses on recursive approaches to least squares filtering, the interested reader is referred to the many excellent texts [5, 6, 8] that contain detailed information on block-oriented approaches.

One very active research area within least squares filtering is that of reduced complexity filters. This is a very broad category that examines alternative recursive approaches requiring less computation and data storage than the general RLS filter of this chapter. The general RLS filter requires on the order of N^2 operations per iteration to implement the N-length least squares prediction filter [1]. This is largely due to the complexity of recursively computing the inverse of a sample covariance matrix. However, Ljung, et al. [9] and Falconer and Ljung [2] exploited a structure inherent in the sample covariance matrix to reduce this number of computations to approximately $10\,N$. This approach was called the "fast Kalman" method and has been applied to many areas, such as channel equalization [2] and echo cancellation [3]. However, recently, the use of the fast Kalman method has been surplanted by even more efficient filters derived using vector space approaches [10–12]. These methods reduce the required operations to approximately $7\,N$ for transversal least squares prediction filtering. Since these vector space approaches are the subjects of Chapters 9–11 of this text, further discussion of those techniques will be deferred until those chapters.

However, prerequisite to an effective use of any of these advanced methods is an understanding of the general RLS problem, which is the subject of this chapter. The RLS prediction problem is developed in Section 8.2 and the optimal recursive filter is computed. Section 8.3 next discusses the issue of computational complexity and compares the efficiency of RLS versus block-oriented approaches. A set of notation is also developed in this chapter that is consistent with the earlier statistical approaches, as well as the modern vector space approaches of Chapters 9–11.

8.2 The Recursive Least Squares Filter

The problem of recursive least squares (RLS) filtering may be described in the following context. Suppose the desired signal is $d(n)$ and the acquired data on which a prediction is to be based is the signal $x(n)$. As in previous chapters, let the linear prediction filter, $\mathbf{w}_N(n)$, be the N-length predictor computed at time n. However, in the least squares case, the error made in predicting the ith desired sample $d(i)$ using $\mathbf{w}_N(n)$ is considered. This error is given the symbol $e(i|n)$ and is defined by

$$e(i|n) = d(i) - \mathbf{x}_N^T(i)\mathbf{w}_N(n). \tag{8.2.1}$$

The reason for this notation will become clear directly. In (8.2.1), $d(i)$ is the desired signal at time i, $\mathbf{x}_N(i)$ is the N-length data vector used in the prediction of $d(i)$,

$$\mathbf{x}_N(n) = [x(i), x(i-1), \ldots, x(i-N+1)]^T, \tag{8.2.2}$$

and $\mathbf{w}_N(n)$ is the N-length vector

$$\mathbf{w}_N(n) = [w_1(n), w_2(n), \ldots, w_N(n)]^T. \tag{8.2.3}$$

The RLS problem requires finding the set of predictor coefficients $\mathbf{w}_N(n)$ such that the *cumulative squared error measure*

$$\varepsilon(n) = \sum_{i=1}^{n} \lambda^{n-i} e^2(i|n) \tag{8.2.4}$$

is minimized. The optimal setting for the predictor $\mathbf{w}_N(n)$ enters (8.2.4) via the definition of the error $e(i|n)$ in (8.2.1). The parameter λ, where $0 < \lambda \leq 1$, is a data-weighting factor that may be used to weight recent data more heavily in the RLS computations. Although λ is very important in the application of RLS to nonstationary data, it is often set equal to 1 if it is known that the data is stationary. A value of $\lambda = 1$ signifies that all data is equally weighted in the computation of $\mathbf{w}_N(n)$, and this case is often referred to in the literature as the *prewindowed* case. Progressively smaller values of λ compute $\mathbf{w}_N(n)$ based upon effectively smaller windows of data that are beneficial in non-stationary environments. For these reasons, λ is sometimes called the fade factor or forgetting factor for the filter. A value of λ in the range $0.95 < \lambda < 0.9995$ has proved to be effective in tracking local nonstationaries. In general, Eleftheriou and Falconer [13] have shown that there is an optimal λ for a given set of data. However, for most cases, the specific choice of λ within the range above is not critical.

At first glance, the error measure in (8.2.4) might not seem to be greatly different from the mean square prediction error used in previous chapters. However, there is actually a subtle but consequently dramatic difference between the two measures. The mean square error (MSE), as represented by (2.2.3), is a statistical measure derived from considering the long-term statistical properties of the signals under consideration. The normal equations (2.3.5) that resulted were therefore also statistical in nature, and these equations were then solved by recursive techniques to obtain the adaptive algorithms. It was seen in Chapter 5 that an approximation to the method of steepest descent led to the popular LMS adaptive algorithm, and in Chapter 7 that an approximation to Durbin's method led to the gradient adaptive lattice.

The important recognition in these approaches is that the MSE measure derived in Chapter 2 is actually not a function of the data acquired by the processor, but instead depends upon the statistical characterization of the data. However, it is immediately seen that the cumulative squared error criterion of (8.2.4) is indeed a function of the actual data vectors $\mathbf{x}_N(n)$, $\mathbf{x}_N(n-1), \ldots, \mathbf{x}_N(1)$, which will be used in the prediction. Moreover, note that the upper limit of the summation in (8.2.4) is the current time sample n, which means that a new error criterion is recomputed and then reoptimized *for every time iteration n*. This is the very powerful property that the least squares predictor satisfies; namely, that the optimal prediction filter for the exact data signals acquired is computed for every time iteration. An intuitively appealing geometrical interpretation of this property will be examined in Chapter 9. Filters with this property have sometimes been called *exact* least squares filters to signify they have been computed from the actual data acquired.

Therefore, consideration of RLS techniques provides prediction filters that are exactly optimal for the acquired data, rather than statistically optimal for a class of data. Consistent with this quite powerful capability, it might be expected that the RLS algorithms are more complex than the previously considered LMS algorithm. If no consideration of the sequential nature of the recursive predictor were made, then this would indeed be the case. However, recent approaches to RLS filtering have taken advantage of certain data properties that allow a substantial reduction in computation, such that the application of real-time RLS signal processing is now feasible. One example of this type of approach is the fast Kalman algorithm [2, 9]. Another that is extremely important for fast adaptive filtering is the family of vector space methods, which are developed in detail in Chapters 9–11. The current chapter develops the general RLS filtering problem, which is an indispensable background for research in the entire area.

The following derivation of the general RLS algorithm will strive for consistency with the notation used thus far in the text. The interested reader is also referred to the excellent treatments by Proakis [6] and Hsia [5] for additional viewpoints on the derivation. The approach is similar to the vector minimization problem first posed in Chapter 2, except that now the cumulative squared error of (8.2.4) is to be minimized with respect to the $\mathbf{w}_N(n)$ at each time n. Therefore, set the gradient of (8.2.4) to zero,

$$\frac{\partial}{\partial \mathbf{w}_N(n)} \varepsilon(n) = 0. \tag{8.2.5}$$

Using (8.2.4), it is easy to show that solution of (8.2.5) for the $\mathbf{w}_N(n)$ gives the equation

$$\mathbf{R}_{NN}(n)\mathbf{w}_N(n) = \mathbf{p}_N(n), \tag{8.2.6}$$

where $\mathbf{R}_{NN}(n)$ is defined now as the sample autocorrelation matrix

$$\mathbf{R}_{NN}(n) = \sum_{i=1}^{n} \lambda^{n-i} \mathbf{x}_N(i) \mathbf{x}_N^T(i), \tag{8.2.7}$$

and the sample cross-correlation vector is given by

$$\mathbf{p}_N(n) = \sum_{i=1}^{n} \lambda^{n-i} d(i) \mathbf{x}_N(i). \tag{8.2.8}$$

The matrix $\mathbf{R}_{NN}(n)$ is called the sample autocorreltion matrix, since for stationary, ergodic data and $\lambda = 1$

$$\lim_{n \to \infty} \frac{1}{n} \mathbf{R}_{NN}(n) = \mathbf{R}_{NN},$$

where $\mathbf{R}_{NN}$ is the true autocorrelation matrix from (2.2.16). The time argument n in $\mathbf{R}_{NN}(n)$ will be used to differentiate between the sample autocorrelation matrix and the true autocorrelation matrix.

The form of the $N \times N$ matrix equation (8.2.6) is very similar to the statistical normal equations as given by (2.3.5). Theoretically it would be possible to compute the inverse matrix $\mathbf{R}_{NN}^{-1}(n)$ and obtain

$$\mathbf{w}_N(n) = \mathbf{R}_{NN}^{-1}(n)\mathbf{p}_N(n). \tag{8.2.9}$$

However, the massive amount of computation required to do this matrix inverse for every time value n prohibits this straightforward approach. The direct matrix inverse requires on the order of N^3 operations and N^3 storage locations to invert the $N \times N$ matrix [16]. More efficient methods for solving the linear equations (8.2.6) still require on the order of $N^3/3$ operations.

Whereas (8.2.9) is rarely directly implemented in practice, the result is very valuable from a conceptual standpoint and leads directly to the RLS algorithm. Since (8.2.6) and, hence, (8.2.9) are valid for any general value of the time variable, they are valid at time $n - 1$ and thus

$$\mathbf{w}_N(n-1) = \mathbf{R}_{NN}^{-1}(n-1)\mathbf{p}_N(n-1). \tag{8.2.10}$$

For the present, assume that the quantities at the previous time $n - 1$ in (8.2.10) are known. This is similar to the assumption in the lattice derivation of Chapter 3 that solutions at previous orders of the lattice were known. It is desired to compute the next set of least squares filter coefficients $\mathbf{w}_N(n)$ from a knowledge of the quantities in (8.2.10) plus the new time data $x(n)$ and $d(n)$, which have been acquired at time n. This is known as the *time-recursive* approach. If this assumption is followed back to the starting time $n = 0$, then this case is equivalent to assuming the existence of an initial inverse $\mathbf{R}_{NN}^{-1}(0)$, plus a recursion to compute $\mathbf{R}_{NN}^{-1}(n)$ from $\mathbf{R}_{NN}^{-1}(n-1)$. At $n = 0$, no data has yet been acquired, and therefore some initial $\mathbf{R}_{NN}^{-1}(n)$ must be assumed. Both Godard [14] and Falconer and Ljung [2] have shown that, within rather large bounds, the choice of $\mathbf{R}_{NN}^{-1}(0)$ is not crucial to the convergence of the RLS algorithm. General guidelines are that $\mathbf{R}_{NN}(0)$ must be nonsingular such that $\mathbf{R}_{NN}^{-1}(0)$ exists.

The other requirement of a time recursion for $\mathbf{R}_{NN}(n)$ is easily derived by expanding (8.2.7):

$$\mathbf{R}_{NN}(n) = \sum_{i=1}^{n-1} \lambda^{n-i}\mathbf{x}_N(i)\mathbf{x}_N^T(i) + \mathbf{x}_N(n)\mathbf{x}_N^T(n),$$

or

$$\mathbf{R}_{NN}(n) = \lambda\mathbf{R}_{NN}(n-1) + \mathbf{x}_N(n)\mathbf{x}_N^T(n). \tag{8.2.11}$$

This is the desired recursion for $\mathbf{R}_{NN}(n)$ in terms of $\mathbf{R}_{NN}(n-1)$. Note that the second term on the right-hand side of (8.2.11) is also an $N \times N$ matrix Therefore, the *matrix inverse lemma* [1, 18] may be used to derive a recursive update for $\mathbf{R}_{NN}^{-1}(n)$ in terms of the previous inverse, $\mathbf{R}_{NN}^{-1}(n-1)$, and the new data acquired at time n. The matrix inverse lemma states that if a square matrix $\mathbf{H}_{NN}$ can be written as

$$\mathbf{H}_{NN} = \mathbf{A}_{NN} + \mathbf{b}_N\mathbf{c}_N^T, \tag{8.2.12}$$

then its inverse $\mathbf{H}_{NN}^{-1}$ may be written as

$$\mathbf{H}_{NN}^{-1} = \mathbf{A}_{NN}^{-1} - \mathbf{A}_{NN}^{-1}\mathbf{b}_N(1 + \mathbf{c}_N^T\mathbf{A}_{NN}^{-1}\mathbf{b}_N)\mathbf{c}_N^T\mathbf{A}_{NN}^{-1}. \tag{8.2.13}$$

(The symbols $\mathbf{A}_{NN}$, $\mathbf{b}_N$, and $\mathbf{c}_N$ merely represent general matrices and should not be confused with any data or filter quantities defined by similar symbols.)

Equation (8.2.11) may then be written in the form (8.2.12) by recognizing the following associations:

$$\mathbf{A}_{NN} = \lambda\mathbf{R}_{NN}(n-1), \tag{8.2.14a}$$

$$\mathbf{b}_N = \mathbf{x}_N(n), \tag{8.2.14b}$$

$$\mathbf{c}_N^T = \mathbf{x}_N^T(n). \tag{8.2.14c}$$

Using these definitions, the inverse $\mathbf{R}_{NN}^{-1}(n)$ becomes (after some intermediate algebra):

$$\mathbf{R}_{NN}^{-1}(n) = \frac{1}{\lambda}\left[\mathbf{R}_{NN}^{-1}(n-1) - \frac{\mathbf{R}_{NN}^{-1}(n-1)\mathbf{x}_N(n)\mathbf{x}_N^T(n)\mathbf{R}_{NN}^{-1}(n-1)}{\lambda + \mu(n)} \right], \tag{8.2.15}$$

where $\mu(n)$ is the scalar

$$\mu(n) = \mathbf{x}_N^T(n)\mathbf{R}_{NN}^{-1}(n-1)\mathbf{x}_N(n). \tag{8.2.16}$$

Notice from (8.2.15) that if it is assumed that $\mathbf{R}_{NN}^{-1}(n-1)$ is known, then all information needed to compute $\mathbf{R}_{NN}^{-1}(n)$ is available when the processor receives $\mathbf{x}_N(n)$.

Two additional definitions will further simplify the work to follow, as well as aid in the conceptual development of the solution. Define the $N \times N$ matrix $\mathbf{C}_{NN}(n)$ as

$$\mathbf{C}_{NN}(n) = \mathbf{R}_{NN}^{-1}(n), \tag{8.2.17}$$

and the $N \times 1$ "gain" vector $\mathbf{g}_N(n)$ as

$$\mathbf{g}_N(n) = \frac{\mathbf{C}_{NN}(n-1)\mathbf{x}_N(n)}{\lambda + \mu(n)}. \tag{8.2.18}$$

The reason for using the terminology "gain" will become apparant soon. Using these definitions, the recursion in (8.2.15) can now be now be written as

$$\mathbf{C}_{NN}(n) = \frac{1}{\lambda}[\mathbf{C}_{NN}(n-1) - \mathbf{g}_N(n)\mathbf{x}_N^T(n)\mathbf{C}_{NN}(n-1)]. \tag{8.2.19}$$

The recursion in (8.2.19) therefore allows the computation of the $N \times N$ matrix inverse $\mathbf{R}_{NN}^{-1}(n)$ in a recursive fashion, rather than requiring an explicit inversion of the $\mathbf{R}_{NN}(n)$ matrix for each iteration n. This technique provides a substantial savings in computation versus the direct matrix inverse, which requires on the order of N^3 (expressed as $O(N^3)$) operations for each inverse [16, 17]. It will be shown in Table 8.2 that the recursion in (8.2.19) only requires $O(N^2)$ operations.

An alternative definition of $\mathbf{g}_N(n)$ may be derived from (8.2.19), which will be very important in the work to follow. It is easily shown that postmultiplying each side of (8.2.19) by $\lambda\mathbf{x}_N(n)$ and simplifying leads to the important relations

$$\mathbf{g}_N(n) = \mathbf{R}_{NN}^{-1}(n)\mathbf{x}_N(n) = \mathbf{C}_{NN}(n)\mathbf{x}_N(n). \tag{8.2.20}$$

One use of this form is in the derivation for the RLS predictor update. Using (8.2.8), it is easy to expand $\mathbf{p}_N(n)$ as follows:

$$\mathbf{p}_N(n) = \lambda\mathbf{p}_N(n-1) + d(n)\mathbf{x}_N(n). \tag{8.2.21}$$

Using (8.2.17), (8.2.19), and (8.2.21) in (8.2.9) then gives

$$\mathbf{w}_N(n) = \frac{1}{\lambda}[\mathbf{C}_{NN}(n-1) - \mathbf{g}_N(n)\mathbf{x}_N^T(n)\mathbf{C}_{NN}(n-1)] \times [\lambda\mathbf{p}_N(n-1) + d(n)\mathbf{x}_N(n)]$$

$$= \mathbf{C}_{NN}(n-1)\mathbf{p}_N(n-1) - \mathbf{g}_N(n)\mathbf{x}_N^T(n)\mathbf{C}_{NN}(n-1)\mathbf{p}_N(n-1)$$

$$+ \frac{1}{\lambda}[\mathbf{C}_{NN}(n-1)\mathbf{x}_N(n) - \mathbf{g}_N(n)\mathbf{x}_N^T(n)\mathbf{C}_{NN}(n-1)\mathbf{x}_N(n)]d(n),$$

and use of (8.2.8) gives

$$\mathbf{w}_N(n) = \mathbf{w}_N(n-1) - \mathbf{g}_N(n)\mathbf{x}_N^T(n)\mathbf{w}_N(n-1)$$

$$+ \frac{1}{\lambda}[\lambda\mathbf{g}_N(n) + \mu(n)\mathbf{g}_N(n) - \mathbf{g}_N(n)\mu(n)]d(n),$$

or finally

$$\mathbf{w}_N(n) = \mathbf{w}_N(n-1) + \mathbf{g}_N(n)[d(n) - \mathbf{x}_N^T(n)\mathbf{w}_N(n-1)]. \qquad (8.2.22)$$

Lastly, (8.1.1) may be used to simplify (8.2.22) and obtain

$$\mathbf{w}_N(n) = \mathbf{w}_N(n-1) + \mathbf{g}_N(n)e(n|n-1). \qquad (8.2.23)$$

Equation (8.2.23) is the desired result for the update recursion for the RLS prediction filter $\mathbf{w}_N(n)$. The complete set of equations necessary for the RLS algorithm are listed in Table 8.1, which also contains initialization information. Despite all the complexity of the RLS algorithm, its purpose is simply to compute the N components of the new gain vector $\mathbf{g}_N(n)$, which are then used in updating the "old" $\mathbf{w}_N(n-1)$ to the "new" $\mathbf{w}_N(n)$ in an optimal least

Table 8.1 Operation Flow for Regular RLS Algorithm

INITIALIZATION:
At $n = 0$:

$$\mathbf{w}_N(0) = \mathbf{x}_N(0) = \mathbf{0}_N \qquad (8.2.24a)$$
$$\mathbf{C}_{NN}(0) = \delta\mathbf{I}_{NN} \ (\delta \gg 1). \qquad (8.2.24b)$$

OPERATION:
For $n = 1$ to $n = final$ do:
(1) Acquire $d(n)$, $\mathbf{x}_N(n)$.
(2) Form prediction error from (8.2.1):

$$e(n|n-1) = d(n) - \mathbf{x}_N^T(n)\mathbf{w}_N(n-1). \qquad (8.2.24c)$$

(3) Calculate new gain vector from (8.2.16) and (8.2.18):

$$\mu(n) = \mathbf{x}_N^T(n)\mathbf{C}_{NN}(n-1)\mathbf{x}_N(n), \qquad (8.2.24d)$$

$$\mathbf{g}_N(n) = \frac{\mathbf{C}_{NN}(n-1)\mathbf{x}_N(n)}{\lambda + \mu(n)}. \qquad (8.2.24e)$$

(4) Update RLS predictor from (8.23):

$$\mathbf{w}_N(\mathrm{n}) = \mathbf{w}_N(n-1) + \mathbf{g}_N(n)e(n|n-1). \qquad (8.2.24f)$$

(5) Update matrix inverse for the next time iteration:

$$\mathbf{C}_{NN}(n) = \frac{1}{\lambda}[\mathbf{C}_{NN}(n-1) - \mathbf{g}_N(n)\mathbf{x}_N^T(n)\mathbf{C}_{NN}(n-1)]. \qquad (8.2.24g)$$

Loop Complete
Return to (1).

---

squares fashion. Note the similarity between the RLS update (8.2.23) and the LMS update (5.3.2) repeated here for comparison and incorporating the current notation:

$$\mathbf{w}_N(n) = \mathbf{w}_N(n - 1) + \alpha e(n|n - 1)\mathbf{x}_N(n). \qquad (5.3.2)$$

Both methods incorporate the latest scalar prediction error in the update term, but they differ substantially in the vector portion of the update term. The LMS algorithm simply uses the contents of the tapped delay line at time n, whereas the RLS method utilizes a rather sophisticated process to compute its vector update.

The benefit of expending this extra effort on the RLS method is an algorithm that converges much faster than the simple LMS method. This will be illustrated in Chapters 10 and 11, which will display results of using the family of least squares prediction methods developed in this text. The qualitative reason for the superiority of RLS is that the LMS algorithm computes the minimum MSE predictor only in the asymptotic case; that is, at convergence of the LMS filter as $n \to \infty$. On the other hand, the RLS algorithm finds the optimal predictor at each point in time by minimizing an exact error measure constructed from the actual data acquired. Moreover, the least squares error measure itself changes at each time iteration n to reflect the acquisition of new data. The gain $\mathbf{g}_N(n)$ is the result of this exact minimization process. Once $\mathbf{g}_N(n)$ has been found, then the old $\mathbf{w}_N(n - 1)$ may be updated according to (8.2.23).

8.3 Computational Complexity

The RLS algorithm developed in the previous section was straightforward in both derivation and the flow of equations (8.2.24). For application of non real-time (or off-time) recursive least squares filtering, the equations in (8.2.24) may indeed be implemented in exactly this manner. However, for the many implementations of recursive least squares filtering requiring real-time operation, the computational cost of the regular RLS algorithm can become prohibitive, especially for large filter lengths N. Even though the RLS method is a substantial improvement over the direct matrix inverse method, there are still many cases in which current processors will not have the computational power or storage capability to implement the regular RLS directly.

To see more clearly the source of the computational complexity, consider Table 8.2, which denotes the number of arithmetic operations associated with each segment of the RLS algorithm. The approach followed in Table 8.2 is extremely useful and will be used in Chapters 10 and 11 as well to calculate the computational counts associated with fast versions of least squares adaptive filters. In the calculations of Table 8.2, the symbols β and γ are simply dummy variables for signifying scalar (β, γ), vector $(\boldsymbol{\beta}_N, \boldsymbol{\gamma}_N)$, or matrix $(\boldsymbol{\beta}_{NN}, \boldsymbol{\gamma}_{NN})$ intermediate computations. The entries shown in the "Mult/Divide" column

Table 8.2 Computational Complexity of Regular RLS Algorithm

Equation	Calculation	Mult/Divide	Add/Subtract
(8.2.24c)	$\gamma = \mathbf{x}_N^T(n)\mathbf{w}_N(n-1)$	N	$N-1$
(8.2.24d)	$\gamma_N = \mathbf{C}_{NN}(n-1)\mathbf{x}_N(n)$	N^2	$N(N-1)$
	$\mu(n) + \mathbf{x}_N^T(n)\gamma_N$	N	N
(8.2.24e):	$\gamma = \lambda + \mu(n)$		1
	$\gamma_N + \mathbf{C}_{NN}(n-1)\mathbf{x}_N(n)$		(No ops., saved from 8.2.24d)
	$\mathbf{g}_N(n) = \gamma_N/\gamma$	N	
(8.2.24f)	$\gamma_N = \mathbf{g}_N(n)e(n|n-1)$	N	
	$\mathbf{w}_N(n) = \mathbf{w}_N(n-1) + \gamma_N$		N
(8.2.24g)	$\gamma_N^T = \mathbf{x}_N^T(n)\mathbf{C}_{NN}(n-1)$	N^2	$N(N-1)$
	$\boldsymbol{\beta}_{NN} = \mathbf{g}_N(n)\gamma_N^T$	N^2	
	$\gamma_{NN} = \mathbf{C}_{NN}(n-1) - \boldsymbol{\beta}_{NN}$		N^2
	$\mathbf{C}_{NN}(n) = \gamma_{NN}/\lambda$	N^2	
	Total:	$4N^2 + 4N$	$3N^2 + N - 1$

are the number of multiplications and/or divisions required to carry out the operation in the "Calculation" column. For example, the computation

$$\gamma = \mathbf{x}_N^T(n)\mathbf{w}_N(n-1)$$

requires N multiplications plus $N-1$ additions, as may be verified by expanding the inner product:

$$\gamma = \mathbf{x}_N^T(n)\mathbf{w}_N(n-1)$$

$$= \sum_{i=1}^{N} x(n-i)w_{N,i}(n-1)$$

$$= x(n-1)w_{N,1}(n-1) + x(n-2)w_{N,2}(n-1) + \cdots + x(n-N)w_{N,N}(n-1).$$

Other operational counts may be verified in a similar fashion.

From Table 8.2, it is seen that approximately $7N^2$ arithmetic operations (additions, subtractions, multiplications, and divisions) are needed per time iteration. However, there are several methods whereby the number of operations may be reduced. One is the fast Kalman algorithm, which is capable of reducing the arithmetic operations to approximately $10N$ per time update. This algorithm is derived and described in detail in [2, 9], and the reader is referred to those works for details. However, a very elegant mathematical method using vector space concepts also reduces the complexity to this level, while retaining a very simple geometrical interpretation for the resulting least squares computations. The concepts of linear vector spaces may be employed to produce the least squares lattice filter and the fast transversal filter, which are currently very active research areas in adaptive signal processing. These techniques, and other geometrical properties of least squares filters, are explored in Chapters 9–11.

PROBLEMS

1. Show that minimization of the error criterion $\varepsilon(n)$ given in (8.2.4) leads to the sampled data form for the "normal equations" given in (8.2.6).

2. Suppose now a new error measure

$$\varepsilon_a(n) = \sum_{i=1}^{n} \lambda^{n-i} e^2(i|n) + \lambda^n \mathbf{w}_N^T(n) \mathbf{w}_N(n)$$

is created, where $\mathbf{w}_N(n)$ is the least squares prediction filter. With $e(i|n)$ defined by (8.2.1), find the equation for the optimal (least squares) filter $\mathbf{w}_N(n)$, which minimizes $\varepsilon_a(n)$ at every n. When does this solution (using $\varepsilon_a(n)$) approach the solution for (8.2.6)?

3. Assume that all data samples $x(n) = 0$ for $n < 1$ and the error measure in (8.2.4) is to be minimized.
   (a) Under these conditions, what happens to the $\mathbf{R}_{NN}(n)$ matrix in (8.2.7) for time $n < N$? Justify your answer.
   (b) What is the implication concerning the theoretical solution (8.2.9)?
   (c) What is a potential remedy for this problem so that $\mathbf{w}_N(n)$ for $n < N$ may be computed?

4. Verify that any matrix $\mathbf{H}_{NN}$, which can be written as (8.2.12), does have the inverse given by (8.2.13).

5. What is the extension of the matrix inverse lemma to find $\mathbf{H}_{NN}$ if

$$\mathbf{H}_{NN} = \mathbf{A}_{NN} + \mathbf{b}_N \mathbf{x}_N^T \mathbf{Q}_{NN} \mathbf{y}_N \mathbf{c}_N^T,$$

where $\mathbf{x}_N$, $\mathbf{y}_N$ are general N dimension vectors and $\mathbf{Q}_{NN}$ is an $N \times N$ matrix.

6. The effect of a data sample $x(n_0)$ upon the RLS computations decreases for time $n > n_0$. Let the time constant τ be defined as the number of samples it takes for $x(n_0)$ to be weighted by a factor of e^{-1} in the computation of $\mathbf{R}_{NN}(n)$. For the simple scalar case $N = 1$, show that $\tau \approx 1/(1 - \lambda)$ for values of $\lambda \approx 1$.

7. One justification for using RLS algorithms is that their computations and storage requirements are less than direct solutions of the linear equations (8.2.6). Assume that the popular linear algebra technique of *gaussian elimination* is used to solve (8.2.6), repeated below, for the $\mathbf{w}_N(n)$:

$$\mathbf{R}_{NN}(n) \mathbf{w}_N(n) = \mathbf{p}_N(n).$$

Also assume for this problem that the elements of $\mathbf{R}_{NN}(n)$ and $\mathbf{p}_N(n)$ are given beforehand and require no computation themselves.
   (a) Compute the number of arithmetic operations (multiplications, adds, divides) required to solve (8.2.6) using gaussian elimination.
   (b) For what size N does it become feasible to use RLS rather than gaussian elimination?

REFERENCES

1. J.G. Proakis, *Digital Communications*, McGraw-Hill, New York, 1983.
2. D.D. Falconer and L. Ljung, "Application of Fast Kalman Estimation to

Adaptive Estimation," *IEEE Trans. on Communications*, vol. COM-26, pp. 1439–1445, October 1978.

3. S.T. Alexander and S.H. Ardalan, *Conf. Record*, 1985 Int. Conf. on Communications, Chicago, IL, June 1985.

4. R.E. Kalman, "A New Approach to Linear Filtering and Prediction Problems," *J. Basic Engineering*, vol. 82, pp. 34–45, March 1960.

5. T.S. Hsia, *Systems Identification: Least Squares Methods*, Lexington Press, Lexington, MA, 1977.

6. G.C. Goodwin and R.L. Payne, *Dynamic System Identification: Experiment Design and Data Analysis*, Academic Press, New York, 1977.

7. K.F. Gauss, "Theora Motus Corporum Coelestium," 1809, reprinted in *Theory of Motion of Heavenly Bodies*, Dover Press, New York, 1963.

8. C.L. Lawson and R.J. Hanson, *Solving Least Squares Problems*, Prentice-Hall, Englewood Cliffs, NJ, 1974.

9. L. Ljung, M. Morf, and D.D. Falconer, "Fast Calculation of Gain Matrices for Recursive Estimation Schemes", *Int. J. of Control*, pp. 1–17, January 1978.

10. J.M. Cioffi, "Fast Transversal Filters for Communications Applications," Ph. D. Dissertation, Stanford University, Stanford, CA, 1984.

11. J.M. Cioffi and T. Kailath, "Fast Recursive-Least-Squares Filters for Adaptive Filtering," *IEEE Trans. on Acous., Speech, and Signal Processing*, vol. ASSP-32, pp. 304–338, April 1984.

12. M.L. Honig, "Recursive, Fixed-Order Covariance Least Squares Algorithms," *Bell Sys. Tech. J.*, vol. 62, pp. 2961–2992, December 1983.

13. E. Eleftheriou and D.D. Falconer, "Steady State Behavior of RLS Adaptive Algorithms," *Proceedings*, 1985 IEEE Int. Conf. on Acous., Speech and Signal Proc., Tampa, FL, March 1985.

14. D. Godard, "Channel Equalization Using a Kalman Filter for Fast Data Transmission," *IBM J. Res. and Dev.*, pp. 267–273, May 1974.

15. Y. Miyanaga, N. Miki, N. Nagai, and K. Hatori, "A Speech Analysis Algorithm which Eliminates the Influence of Pitch Using the Model Reference Adaptive System," *IEEE Trans. on Acoustics, Speech and Signal Processing*, vol. ASSP-30, pp. 88–95, February 1982.

16. B. Noble, *Applied Linear Algebra*, Prentice-Hall, Englewood Cliffs, NJ, 1969.

17. R. Bellman, *Introduction to Matrix Analysis*, McGraw-Hill, New York, 1960.

18. G.J. Bierman, *Factorization Methods for Discrete Sequential Estimation*, Academic Press, New York, 1977.

CHAPTER 9
Vector Spaces for RLS Filters

9.1 Introduction

This chapter introduces a very important and mathematically elegant approach to recursive least squares adaptive filtering. This approach employs the very powerful mathematics of linear vector spaces. Through the use of vector space concepts, many intuitive geometrical interpretations result that will enable the reader to understand many approaches currently used in the literature. Both the least squares lattice (LSL) and the least squares transversal filter may be derived once this vector space interpretation has been obtained. The LSL is derived in detail in Chapter 10 and the fast transversal filter is the subject of Chapter 11. The current chapter introduces and develops applicable vector space relations and geometrical concepts that are necessary for either fast filter derivation.

This chapter is organized as follows. Section 9.2 provides an introduction to necessary background in linear vector spaces, and Section 9.3 then explores the relation between vector space concepts, projection matrices, and the least squares prediction. Section 9.4 uses these results to next derive the general least squares update equations for a vector space that is changing in a specified manner. Finally, Section 9.5 then develops the specific case of the least squares time-update equations in a general form, such that it may be immediately used in the lattice development of Chapter 10 and the transversal filter development of Chapter 11.

9.2 Linear Vector Spaces

The development and discussion of projection operators and vector spaces to follow will provide the mathematical tools for the lattice filter derivation of Chapter 10 and the transversal filter of Chapter 11. Much of the background of work to come expands upon classical linear algebra results [1–3] by extending certain concepts to the signal processing problems of order- and time-recursive updating. In general, the study of vector spaces is a rich and widely applicable discipline, and the current discussion will only utilize a small portion of the subject. An excellent introductory text is that by Luenberger [4], which treats the most general form of the problem; namely, an optimization in a Hilbert space. At a more advanced level, the classical text by Halmos [5] examines the more mathematical aspects of vector spaces. Additionally, the text by DeRusso, Roy, and Close [6] presents an excellent development of basis vectors and vector space transformations in the context of circuits and systems applications. For present applications, the problem is simplified somewhat to that of minimization of matrix norms in a linear vector space. Detailed developments of these topics, as well as other applications, are contained in the texts by Stewart [7] and Lawson and Hanson [8]. Additionally, the paper by Youla [9] provides an advanced application of projection techniques to a problem in signal restoration. The text by Naylor and Sell [10] treats the broader discipline of linear operators, of which projection techniques are a subset. The interested reader is directed to these works for additional material.

The basis for the derivation of fast recursive least squares (RLS) filters is the formulation of a linear vector space. Therefore, at least a working knowledge of linear vector spaces is needed as background. Consider the vector $\mathbf{x}(n)$ of data that has been acquired at time n:

$$\mathbf{x}(n) = [x(1), x(2), \ldots, x(n-1), x(n)]^T. \qquad (9.2.1)$$

With no loss of generality, time has been assumed to start at time, "1" and the current sample is time n. The vector of data samples in (9.2.1) is seen to be have n components, which are illustrated in Figure 9.1 for the case of the signal $\{x(n)\} = \{2, 3, 4, \ldots\}$. For instance, $\mathbf{x}(1)$ would be a vector of amplitude $x(1) = 2$ lying along the first coordinate axis; $\mathbf{x}(2)$ would be a vector $[2, 3]^T$ in the plane spanned by the first and second coordinate axes having individual components $x(1)$ and $x(2)$; and $\mathbf{x}(3)$ would be the three-component vector given by

$$\mathbf{x}(3) = [x(1), x(2), x(3)]^T = [2, 3, 4]^T. \qquad (9.2.2)$$

The importance of vectors of the form $\mathbf{x}(n)$ is primarily in defining a specific *vector space*. For example, even though $\mathbf{x}(n)$ is a vector with n components, it only spans a one-dimensional space, which is a line in the direction of $\mathbf{x}(n)$. Two distinct n component vectors, $\mathbf{x}_1(n)$ and $\mathbf{x}_2(n)$, span a two-dimensional

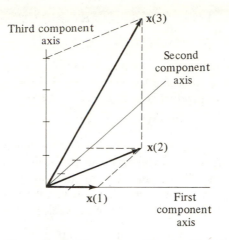

Figure 9.1   Example of $\mathbf{x}(n)$ vectors: $x(1) = 2$, $x(2) = 3$, $x(3) = 4$.

space, which is a plane containing both of the vectors. Any vector $\mathbf{u}(n)$ in this plane may then be written as a linear combination of $\mathbf{x}_1(n)$ and $\mathbf{x}_2(n)$:

$$\mathbf{u}(n) = a_1 \mathbf{x}_1(n) + a_2 \mathbf{x}_2(n), \tag{9.2.3}$$

where a_1 and a_2 are constants. In fact, (9.2.3) suggests an informal working definition for a linear vector space. For purposes of this book, the m-dimensional linear vector space $\{\mathbf{U}\}$ will be defined as the set of vectors that can be written as a linear combination of the m *basis vectors* of the vector space; that is

$$\{\mathbf{U}\} = \{\mathbf{u}(n)|\mathbf{u}(n) = \sum_{k=1}^{m} a_k \mathbf{x}_k(n)\}, \tag{9.2.4}$$

where the $\mathbf{x}_k(n)$ are the basis vectors. The dimension of the vector space will be the minimum number of basis vectors required to span the space. Additionally, all vector spaces examined in this text have a finite number of basis vectors and are therefore finite dimension vector spaces. For a more rigorous definition of a linear vector space, including allowed operations and examples, the reader is referred to [5, 6, 10].

The "length" of a vector in the vector space may be defined using the vector norm

$$\langle \mathbf{u}(n), \mathbf{u}(n) \rangle^{1/2} = [\mathbf{u}^T(n)\mathbf{u}(n)]^{1/2} = \left[\sum_{k=1}^{n} u^2(k)\right]^{1/2}. \tag{9.2.5}$$

Equation (9.2.5) actually represents a special case of a more general operation called the inner product of two general vectors, $\langle \mathbf{u}(n), \mathbf{v}(n) \rangle$, defined by

$$\langle \mathbf{u}(n), \mathbf{v}(n) \rangle = \mathbf{u}^T(n)\mathbf{v}(n) = \sum_{k=1}^{n} u(k)v(k). \tag{9.2.6}$$

The structure of (9.2.6) may also be used to define a matrix inner product.

Given two matrices $\mathbf{U}$ and $\mathbf{X}$, the inner product $\langle \mathbf{U}, \mathbf{X} \rangle$ will be defined as

$$\langle \mathbf{U}, \mathbf{X} \rangle = \mathbf{U}^T \mathbf{X}. \tag{9.2.7}$$

This definition will hold for any two matrices $\mathbf{U}$ and $\mathbf{X}$ such that the dimensions on $\mathbf{U}^T$ and $\mathbf{X}$ are compatible for matrix multiplication. Another valuable property of the inner product is its linearity, which provides the relation

$$\langle \mathbf{U}, \mathbf{X} + \mathbf{V} \rangle = \langle \mathbf{U}, \mathbf{X} \rangle + \langle \mathbf{U}, \mathbf{V} \rangle. \tag{9.2.8}$$

Using the concept of vector norm, a metric for "measuring" the distance between two vectors $\mathbf{u}(n)$ and $\mathbf{v}(n)$ may now be defined. In general, the form a distance metric takes is limited only by imagination. However, in the current development, the very common metric known as the Euclidean distance [4] will be used. This distance is the norm of the vector $\mathbf{u}(n) - \mathbf{v}(n)$, or

$$d[\mathbf{u}(n), \mathbf{v}(n)] = \langle \mathbf{u}(n) - \mathbf{v}(n), \mathbf{u}(n) - \mathbf{v}(n) \rangle^{1/2}. \tag{9.2.9}$$

A linear vector space having a metric induced by an inner product is called an inner product space. Furthermore, if the inner product space is *complete*, then it is a *Hilbert space*. All the linear vector spaces examined in this book are indeed complete and are therefore Hilbert spaces. Most applications of recursive filtering for signal processing will involve collections of vectors, such that the definitions of (9.2.1)–(9.2.9) involve two vectors or a matrix and a vector.

Subspaces

A concept central to the development of vector space approaches to RLS is that of a subspace of the linear vector space. A familiar geometrical analogy from Euclidean geometry is the xy-plane being a subspace of the three-dimensional space spanned by the orthogonal x, y, and z axes. Very often a "projection" will be made from a point in an $m + 1$ dimension linear vector space to an m dimension data subspace. It will be developed that the least squares filter is constrained to produce a prediction that can only lie in the data subspace, whereas the vector being predicted usually lies in a higher dimensional space. This vector being predicted will be called the desired vector for reasons to be developed shortly, and usually the desired vector has one more dimension than the subspace. The least squares prediction problem then is very frequently locating the position in the m dimension data subspace that is "nearest," in some sense, to the desired vector (which is in the $m + 1$ dimensional space).

The projection method approaches the solution by using the actual acquired data to create a set of basis vectors that span the m dimension data subspace. Then, an operator called the projection matrix premultiplies the current desired vector to produce a projection of the desired vector onto the data subspace. This projected vector will be the least squares estimate; that is, it is the vector nearest (in a least squares sense) to the desired vector, but which lies entirely within the data subspace. It should be stressed that the projection matrix provides the concept for the derivation of the fast RLS algorithms. However, the fast filter structures that result are not physically implemented by actually computing matrix multiplications. The projection matrix is a conceptual tool that is invaluable in deriving and understanding the fast RLS algorithms that do result.

9.3 The Least Squares Filter and Projection Matrices

There are several versions of fast algorithms that could be derived using a vector space approach. However, for introductory purposes, the simplest case will most easily illustrate the geometrical principles involved. The form of LSL filter to be developed in Chapter 10 and the transversal filter of Chapter 11 are similar to the case of prewindowed data of Chapter 8, in which the underlying physical processes are assumed to be stationary. The prewindowed assumption implies that all data could be equally weighted in the computations. To determine the particular vector space that is applicable to this case, consider again the computation of each of the n least squares prediction errors from (8.2.4):

$$e(1|n) = d(1) - x(1)w_1(n)$$

$$e(2|n) = d(2) - x(2)w_1(n) - x(1)w_2(n)$$

$$\cdots \tag{9.3.1}$$

$$e(n|n) = d(n) - x(n)w_1(n) - \cdots - x(n - m + 1)w_m(n),$$

where all quantities prior to time 1 have been assumed to be zero. As in Chapter 8, the notation "$e(i|j)$" signifies the error in "predicting $d(i)$ based on using the filter computed at time j." Note that (9.3.1) can be written in matrix form as

$$\mathbf{e}(n|n) = \mathbf{d}(n) - \hat{\mathbf{d}}(n), \tag{9.3.2}$$

where

$$\hat{\mathbf{d}}(n) = \mathbf{X}_{0,m-1}(n)\mathbf{w}_m(n), \tag{9.3.3}$$

and

$$\mathbf{X}_{0,m-1}(n) = \begin{bmatrix} x(1) & 0 & \cdots & 0 \\ x(2) & x(1) & & 0 \\ \cdots & \cdots & & \cdots \\ \cdots & \cdots & & \cdots \\ x(n-1) & x(n-2) & \cdots & x(n-m) \\ x(n) & x(n-1) & \cdots & x(n-m+1) \end{bmatrix}. \tag{9.3.4}$$

The bold lower case notation (without a subscript) followed by a time argument will be reserved to denote vectors having n components, corresponding to the current time index, n. Thus, $\mathbf{x}(n)$ is a vector of n components with final element $x(n)$. By the same token, $\mathbf{x}(n-1)$ is a vector of $n-1$ components with final element $x(n-1)$.

An alternative representation of (9.3.4) will be very valuable in work to come. Note that each of the columns of $\mathbf{X}_{0,m-1}(n)$ is simply a shifted version of $\mathbf{x}(n)$, with the most recent sample values being shifted "off the end" in the latter columns. This naturally suggests the usage of the delay operator z^{-j} once again. Thus, define the time-shifted vector

$$z^{-j}\mathbf{x}(n) = [0, 0, \ldots, 0, x(1), \ldots, x(n-j)]^T. \tag{9.3.5}$$

Note that z^{-j} represents an operator and not a premultiplication. Using this definition, $\mathbf{X}_{0,m-1}(n)$ can then be defined as the matrix given by

$$\mathbf{X}_{0,m-1}(n) = [z^0\mathbf{x}(n), z^{-1}\mathbf{x}(n), \ldots, z^{-(m-1)}\mathbf{x}(n)]. \tag{9.3.6}$$

The subscripts on $\mathbf{X}_{0,m-1}(n)$ correspond to the range of time shifts on the column vectors, and the range of the subscripts denotes the number of columns in the matrix. The time argument gives the number of rows. It is easy to show that the partitioning of (9.3.6) allows (9.3.4) to be rewritten as

$$\hat{\mathbf{d}}(n) = w_1(n)\mathbf{x}(n) + \cdots + w_m(n)z^{-m+1}\mathbf{x}(n), \tag{9.3.7}$$

from which $\hat{\mathbf{d}}(n)$ may be interpreted as a linear combination of the columns of $\mathbf{X}_{0,m-1}(n)$. The vector space interpretation is that the entire vector $\mathbf{d}(n)$ is being predicted in (9.3.7), which is a broader interpretation than the scalar prediction case of Chapters 3, 5, and 7.

Previously in Chapter 8, the sum of the squares of the error components in (9.3.2) was minimized with respect to $\mathbf{w}_m(n)$. However, note that using the definition of inner product from (9.2.5) allows (8.2.4) to be rewritten as

$$\varepsilon(n) = \langle \mathbf{e}(n|n), \mathbf{e}(n|n) \rangle, \tag{9.3.8}$$

where $\lambda = 1$ has been used in (8.2.4) for the prewindowed case. Therefore, the least squares problem of Chapter 8 is equivalent to minimizing the vector norm expressed in (9.3.8). This is the stage at which the transition to a vector space approach begins. In this process, the relation between the error minimizations of the previous chapters and the simple concept of a projection matrix will be derived.

If the previous methods for error square minimization were followed, the equation

$$\frac{\partial}{\partial \mathbf{w}_m(n)} \langle \mathbf{e}(n|n), \mathbf{e}(n|n) \rangle = 0 \tag{9.3.9}$$

would be solved next to find the set of least squares predictor coefficients $\mathbf{w}_m(n)$. Such a procedure would produce a solution

$$\mathbf{w}_m(n) = [\mathbf{X}_{0,m-1}^T(n)\mathbf{X}_{0,m-1}(n)]^{-1}\mathbf{X}_{0,m-1}^T(n)\mathbf{d}(n)$$

$$= \langle \mathbf{X}_{0,m-1}(n), \mathbf{X}_{0,m-1}(n) \rangle^{-1} \langle \mathbf{X}_{0,m-1}(n), \mathbf{d}(n) \rangle. \qquad (9.3.10)$$

Using this result, the prediction $\hat{\mathbf{d}}(n)$ is given from (9.3.3) as

$$\hat{\mathbf{d}}(n) = \mathbf{X}_{0,m-1}(n)\mathbf{w}_m(n)$$

$$= \mathbf{X}_{0,m-1}(n)\langle \mathbf{X}_{0,m-1}(n), \mathbf{X}_{0,m-1}(n) \rangle^{-1}\mathbf{X}_{0,m-1}^T(n)\mathbf{d}(n). \qquad (9.3.11)$$

This is the well-known least squares solution [2, 8, 11]. In deriving (9.3.10), it has been assumed that the matrix product is full-rank, such that the true inverse exists. This will indeed be the case for the vast majority of physical applications, and the current scope of investigations will be limited to this case. In general, even if the matrix product is not full-rank, a solution of (9.3.10) will still exist except that now the simple inverse must be replaced by the pseudoinverse of the matrix product. The interested reader is directed to Noble [12] and Albert [13] for more information on these pseudoinverse problems.

Returning to (9.3.3), the least squares prediction $\hat{\mathbf{d}}(n)$ is seen to be formed by premultiplying $\mathbf{w}_m(n)$ by $\mathbf{X}_{0,m-1}(n)$. Hence, $\hat{\mathbf{d}}(n)$ is a linear combination of the columns of $\mathbf{X}_{0,m-1}(n)$ and therefore lies in the m dimension subspace spanned by the columns of $\mathbf{X}_{0,m-1}(n)$. This particular subspace is called the column space of $\mathbf{X}_{0,m-1}(n)$ or the range of $\mathbf{X}_{0,m-1}(n)$ [2] and will be denoted by the subspace $\{\mathbf{X}_{0,m-1}(n)\}$. In this text, a vector space or subspace will be denoted by placing the appropriate defining matrix in brackets. Therefore, the operation $\mathbf{X}_{0,m-1}(n)\mathbf{w}_m(n)$ displayed in (9.3.11) takes a vector $\mathbf{d}(n)$ in the $m + 1$ dimension linear vector space and produces the prediction $\hat{\mathbf{d}}(n)$, which is a vector in the m dimension subspace $\{\mathbf{X}_{0,m-1}(n)\}$. This subspace is therefore the subspace that contains the least squares prediction of the desired vector.

The preceding can be simplified using the concept of the *projection matrix*. Note that all matrix operations upon $\mathbf{d}(n)$ on the right-hand side of (9.3.11) can be grouped and written as

$$\mathbf{X}_{0,m-1}(n)\mathbf{w}_m(n) = \mathbf{P}_{0,m-1}(n)\mathbf{d}(n),$$

where

$$\mathbf{P}_{0,m-1}(n) = \mathbf{X}_{0,m-1}(n)\langle \mathbf{X}_{0,m-1}(n), \mathbf{X}_{0,m-1}(n) \rangle^{-1}\mathbf{X}_{0,m-1}^T(n). \qquad (9.3.12)$$

The matrix $\mathbf{P}_{0,m-1}(n)$ is a member of a special class of matrices called projection matrices. This concept will be used frequently in work to come, and it will be valuable to understand the relationships between vector spaces, projection matrices, and data matrices. In general, the data matrix $\mathbf{U}$ has columns that span the vector space denoted by $\{\mathbf{U}\}$. The associated projection matrix $\mathbf{P}_{\mathbf{U}}$, which "projects" vectors onto $\{\mathbf{U}\}$, is then defined by

$$\mathbf{P}_{\mathbf{U}} = \mathbf{U}\langle \mathbf{U}, \mathbf{U} \rangle^{-1}\mathbf{U}^T. \qquad (9.3.13)$$

Note that $\mathbf{U}$ may be a matrix or vector of data or even a scalar for that matter.

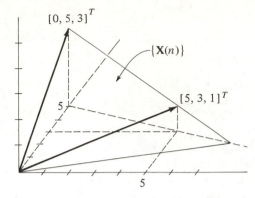

Figure 9.2   Basis vectors $\{\mathbf{x}(n), z^{-1}\mathbf{x}(n)\}$ for data subspace $\{\mathbf{X}(n)\}$. Data signal $\{x(n)\} = \{5, 3, 1, \dots\}$.

Projection matrices have a number of interesting and useful properties that are well developed in [2, 12]. Two properties that will be extremely useful in the current work are

$$\mathbf{P_U P_U} = \mathbf{P_U} \qquad\qquad (9.3.14a)$$

where

$$\mathbf{P_U^T} = \mathbf{P_U}. \qquad\qquad (9.3.14b)$$

Therefore, using the definition of the projection matrix in (9.3.3) immediately gives

$$\hat{\mathbf{d}}(n) = \mathbf{P}_{0, m-1}(n)\mathbf{d}(n), \qquad\qquad (9.3.15)$$

signifying that the least squares prediction is simply the projection of the desired vector onto $\{\mathbf{X}_{0, m-1}(n)\}$.

This has a very pleasing geometrical analogy, as illustrated in Figure 9.2, for an example using $m = 2$, $n = 3$, and the data signal $\{x(n)\} = \{5, 3, 1, \dots\}$. The data subspace $\{\mathbf{X}(n)\}$ is the plane formed by any linear combination of $\mathbf{x}(n) = [5, 3, 1]^T$ and $z^{-1}\mathbf{x}(n) = [0, 5, 3]^T$. The operation in (9.3.15) would form the projection of any vector onto the subspace spanned by the basis vectors $\mathbf{x}(n)$ and $z^{-1}\mathbf{x}(n)$. Note that the basis vectors need not be orthogonal themselves, and that the plane they define may, in general, have any orientation.

The error vector $\mathbf{e}(n|n)$ from (9.3.2) has a special interpretation as well. The scalar error sample $e(n|n)$ is the component of the desired signal $d(n)$ that is unpredictable based upon a knowledge of the m data samples $x(n), \dots$, $x(n - m + 1)$. In the vector space interpretation, the error vector $\mathbf{e}(n|n)$ similarly represents that portion of the desired signal vector that is unpredictable using the m basis vectors of the acquired data subspace. That is, $\mathbf{e}(n|n)$ cannot have a component in $\{\mathbf{X}_{0, m-1}(n)\}$, or, equivalently, the projection of $\mathbf{e}(n|n)$ onto $\{\mathbf{X}_{0, m-1}(n)\}$ must vanish. This only holds if $\mathbf{e}(n|n)$ is a vector normal, or orthogonal, to $\{\mathbf{X}_{0, m-1}(n)\}$.

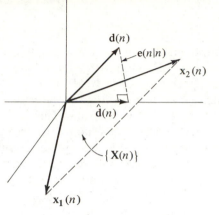

Figure 9.3 Geometrical interpretation of least squares prediction. Data matrix $\mathbf{X}(n) = [\mathbf{x}_1(n), \mathbf{x}_2(n)]$.

Once again, the important concept of orthogonality is demonstrated. In solving the normal equations of Chapter 2, a statistical orthogonality condition was required between the data and the prediction error. Then in Chapter 8, the RLS method enforced an exact orthogonality condition between the prediction error and the data. In the current projection method approach, the concept of orthogonality is present once more and is seen to have a very simple geometrical interpretation. Consider, in Figure 9.3, trying to predict a vector $\mathbf{d}(n)$ that does not lie in the column space of a general matrix $\mathbf{X}(n) = [\mathbf{x}_1(n), \mathbf{x}_2(n)]$. Again, the example is presented for the case of $m = 2$ basis vectors with $n = 3$ components in each vector. The basis vectors $\mathbf{x}_1(n)$ and $\mathbf{x}_2(n)$ span the vector space $\{\mathbf{X}(n)\}$, which is a plane, but $\mathbf{d}(n)$ does not lie in this plane. The least squares prediction, $\hat{\mathbf{d}}(n)$, is the vector in $\{\mathbf{X}(n)\}$ that is closest to $\mathbf{d}(n)$. From simple geometrical considerations, it is easy to see that this vector is the projection of $\mathbf{d}(n)$ onto the plane $\{\mathbf{X}(n)\}$ and, additionally, the prediction error vector $\mathbf{e}(n|n)$ is given by the projection of $\mathbf{d}(n)$ orthogonal to this plane. As stated before and illustrated by Figure 9.3, $\mathbf{e}(n|n)$ does not have a component in $\{\mathbf{X}(n)\}$. Since the concept of a projection orthogonal to a subspace is so important in work to come, this will be investigated in more detail in the next section.

Orthogonal projection matrix

An additional concept needed is that of the orthogonal complement of a projection onto a subspace. The orthogonal complement of a projection is a direct result of the decomposition of a linear vector space into subspaces that are mutually orthogonal. As a working definition, two subspaces, $\{\mathbf{U}\}_p$ and $\{\mathbf{V}\}_q$, will be defined to be orthogonal if for any vector $\mathbf{u}(n)$ in $\{\mathbf{U}\}_p$ and $\mathbf{v}(n)$

in $\{V\}_q$ the relation

$$\langle \mathbf{u}(n), \mathbf{v}(n) \rangle = 0 \qquad (9.3.16)$$

is true. The subscripts on the set notations signify the dimension of each subspace. However, each of $\mathbf{u}(n)$ and $\mathbf{v}(n)$ must have the same number of components for the inner product to be defined. The main benefit of this concept is that it allows a decomposition of the m dimension linear vector space $\{X\}_m$ into a set of orthogonal subspaces. This decomposition will be the key to deriving the LSL updates of Chapter 10 and the transversal least squares updates of Chapter 11. As an example of decompositions, consider the current case of predicting $\mathbf{d}(n)$ using the basis vectors from $\{X_{0,m-1}(n)\}$. As illustrated in the previous discussion, $\mathbf{d}(n)$ is a vector in an $m + 1$ dimension linear vector space. However, $\mathbf{d}(n)$ can be written as a sum of $\hat{\mathbf{d}}(n)$, a vector in the m dimension subspace $\{X_{0,m-1}(n)\}$, plus an error vector $\mathbf{e}(n|n)$ orthogonal to this subspace. All that is needed to define a one-dimensional subspace orthogonal to $\{X_{0,m-1}(n)\}$ is a single basis vector in the orthogonal subspace, since any vector in this subspace may then be written as a scaled version of the basis vector. Since, by definition, the prediction error vector $\mathbf{e}(n|n)$ is orthogonal to $\{X_{0,m-1}(n)\}$, then $\mathbf{e}(n|n)$ will suffice as a basis vector for the subspace orthogonal to $\{X_{0,m-1}(n)\}$. This subspace may be denoted as the error subspace $\{\mathbf{e}(n|n)\}$, and it is a line perpendicular to $\{X_{0,m-1}(n)\}$. The entire $m + 1$ dimensional linear vector space $\{X\}_{m+1}$ containing $\mathbf{d}(n)$ may now be defined as the *direct sum* of the two orthogonal subspaces

$$\{X\}_{m+1} = \{X_{0,m-1}(n)\} + \{\mathbf{e}(n|n)\}. \qquad (9.3.17)$$

This is important, since it introduces a formal structure for decomposing desired signal vectors into their "predictable" components, which lie in the data subspace, and their nonpredictable or error components, which lie in the error subspace. This concept is really the foundation for deriving the fast RLS order and time update relations.

The mathematical tool that produces the projections in these orthogonal subspaces is the orthogonal projection matrix. For example, from (9.3.2) and (9.3.15), the error vector $\mathbf{e}(n|n)$ can be written as

$$\mathbf{e}(n|n) = \mathbf{P}_{0,m-1}^{\perp}(n)\mathbf{d}(n), \qquad (9.3.18)$$

where

$$\mathbf{P}_{0,m-1}^{\perp}(n) = \mathbf{I} - \mathbf{P}_{0,m-1}(n) \qquad (9.3.19)$$

is called the orthogonal projection matrix for the subspace $\{X_{0,m-1}(n)\}$. In (9.3.18), the orthogonal projection matrix operates on $\mathbf{d}(n)$ to produce the component of $\mathbf{d}(n)$ that is orthogonal to $\{X_{0,m-1}(n)\}$. In general, an orthogonal projection matrix $\mathbf{P}_U^{\perp}$ operating on a vector $\mathbf{d}(n)$ will produce the component of $\mathbf{d}(n)$ that is in the overall vector space, but is also orthogonal to $\{U\}$. It can be shown that $\mathbf{P}_U^{\perp}$ has all the beneficial properties of projection matrices previously described for $\mathbf{P_U}$. Let $\mathbf{P_U^{\perp}}$ be defined by

$$\mathbf{P_U^{\perp}} = \mathbf{I} - \mathbf{P_U}. \tag{9.3.20}$$

Then two properties that will be of importance are

$$[\mathbf{P_U^{\perp}}]^T = \mathbf{P_U^{\perp}} \tag{9.3.21}$$

and

$$\mathbf{P_U^{\perp}}\mathbf{P_U^{\perp}} = \mathbf{P_U^{\perp}}. \tag{9.3.22}$$

Another important relation occurs due to the orthogonality between the projection of a general vector $\mathbf{v}$ onto $\{\mathbf{U}\}$ and the projection of $\mathbf{v}$ orthogonal to $\{\mathbf{U}\}$:

$$\langle \mathbf{P_U^{\perp}}\mathbf{v}, \mathbf{P_U}\mathbf{v} \rangle = \mathbf{v}^T \mathbf{P_U^{\perp}} \mathbf{P_U} \mathbf{v} = 0. \tag{9.3.23}$$

Since $\mathbf{v}$ may, in general, be any non-zero vector, then (9.3.23) requires that

$$\mathbf{P_U^{\perp}}\mathbf{P_U} = 0_{nn}, \tag{9.3.24}$$

where 0_{nn} represents the $n \times n$ matrix with all zero elements.

The preceding discussion has introduced projections and orthogonal projections in the familiar context of the least squares prediction error. However, the main importance of projection methods is that they provide a general framework for decomposing vector space elements into their "predictable" and "nonpredictable" components. However, nothing in the current development has thus far suggested an actual implementation of the desired method. Therefore, the results obtained so far are quite general and may be applied to both lattice and transversal implementations of fast RLS filters. The next section develops a general recursion for updating the least squares relations as a vector space "changes" in some specified manner, which will be seen to be applicable to time and order updates for least squares filters.

9.4 Least Squares Update Relations

The least squares lattice and fast transversal filters require updating as new data is made available. The vector space interpretation thus requires the various operators to be updated such that the required least squares orthogonality conditions are maintained. The various forms for the LS matrix, vector, and scalar updates are derived in this section so that they may be applied as needed in succeeding work.

Assume that the current data subspace is $\{\mathbf{U}\}$, which has associated with it the projection matrix $\mathbf{P_U}$ and the orthogonal projection matrix $\mathbf{P_U^{\perp}}$. Now suppose the additional vector $\mathbf{v}$ is added to the set of basis vectors of $\{\mathbf{U}\}$. In the least squares update problem, the specific $\mathbf{v}$ vector added differs depending on the application. However, one obvious application is that of increasing

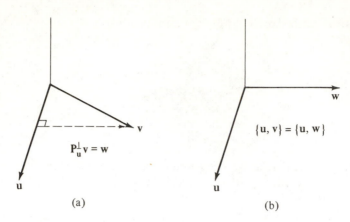

Figure 9.4   Construction of $\{\mathbf{u}, \mathbf{v}\}$ vector space, $\{\mathbf{u}, \mathbf{w}\}$ vector space, and orthogonal basis vectors $\mathbf{u}$, $\mathbf{w}$.

filter order. For example, if $\mathbf{U} = \{\mathbf{X}_{0,m-1}(n)\}$ for an mth order least squares filter, then choosing $\mathbf{v} = z^{-m}\mathbf{x}(n)$ would provide $\{\mathbf{U}, \mathbf{v}\} = \{\mathbf{X}_{0,m}(n)\}$. This is the data subspace for an $m + 1$st order least squares filter. In general, the $\mathbf{v}$ vector will usually provide some information not contained in the basis vectors of $\{\mathbf{U}\}$. As the data subspace changes from $\{\mathbf{U}\}$ to $\{\mathbf{U}, \mathbf{v}\}$, then it is necessary to find the "new" projection matrices $\mathbf{P_{Uv}}$ and $\mathbf{P_{Uv}^{\perp}}$, in terms of the "old" projection matrices $\mathbf{P_U}$ and $\mathbf{P_U^{\perp}}$. If this can be done in a recursive manner, then the groundwork will be laid for the fast RLS filter updates.

These updates are easily done using the concept of orthogonal vector spaces. Note that the vector $\mathbf{v}$ itself cannot be guaranteed to be orthogonal to $\{\mathbf{U}\}$. However, if $\mathbf{v}$ may be used to create a vector $\mathbf{w}$ that is orthogonal to $\{\mathbf{U}\}$, then the projection matrix $\mathbf{P_{U,v}}$ may be decomposed into

$$\mathbf{P_{Uv}} = \mathbf{P_U} + \mathbf{P_w}. \tag{9.4.1}$$

For example, in (9.3.17), $\{\mathbf{U}\}$ would correspond to $\{\mathbf{X}_{0,m-1}(n)\}$, $\{\mathbf{U}, \mathbf{v}\}$ would correspond to $\{\mathbf{X}\}_{m+1}$, and the vector $\mathbf{w}$ would correspond to the $e(n|n)$. Recall that if $\{\mathbf{w}\}$ is one dimensional, then all that is needed to create a basis for $\{\mathbf{w}\}$ is one vector in $\{\mathbf{w}\}$. As shown in Figure 9.4 for the one-dimensional $\{\mathbf{U}\}$ case, the projection of $\mathbf{v}$ orthogonal to $\{\mathbf{U}\}$

$$\mathbf{w} = \mathbf{P_U^{\perp}v} \tag{9.4.2}$$

provides a vector $\mathbf{w}$ orthogonal to $\{\mathbf{U}\}$. The subspace spanned by $[\mathbf{U}, \mathbf{v}]$ is the same as that spanned by $[\mathbf{U}, \mathbf{w}]$, but $\mathbf{U}$ and $\mathbf{w}$ have the added benefit of being orthogonal. The reader familiar with the generation of a set of orthogonal basis vectors will recognize the operations in Figure 9.4 to be a Gram–Schmidt orthogonalization. Therefore, using the definition of a projection matrix given by (9.3.13) in (9.4.1) produces

$$\mathbf{P_{Uv}} = \mathbf{P_U} + \mathbf{P_U^{\perp}v}\langle \mathbf{P_U^{\perp}v}, \mathbf{P_U^{\perp}v}\rangle^{-1}\mathbf{v}^T\mathbf{P_U^{\perp}}. \tag{9.4.3a}$$

This is the update recursion for the new projection matrix onto the space $\{U, v\}$ in terms of the projection matrix onto the space $\{U\}$. The update recursion for the generalized orthogonal projection matrix is then easily seen to be

$$P_{Uv}^{\perp} = I - P_{Uv} = P_U^{\perp} - P_U^{\perp} v \langle P_U^{\perp} v, P_U^{\perp} v \rangle^{-1} v^T P_U^{\perp}. \qquad (9.4.3b)$$

For any general vector y in $\{U, v\}$, equations (9.4.3) may be postmultiplied by y to derive the following orthogonal updates:

$$P_{Uv} y = P_U y + P_U^{\perp} v \langle P_U^{\perp} v, P_U^{\perp} v \rangle^{-1} \langle v, P_U^{\perp} y \rangle, \qquad (9.4.4a)$$

$$P_{Uv}^{\perp} y = P_U^{\perp} y - P_U^{\perp} v \langle P_U^{\perp} v, P_U^{\perp} v \rangle^{-1} \langle v, P_U^{\perp} y \rangle. \qquad (9.4.4b)$$

Note that (9.4.4) gives the update of the entire n component vector. Therefore, this relation will be very useful in order updating entire vectors in least squares recursions. For example, from (9.3.15), it is seen that the update of the predicted vector $\hat{d}(n)$ could be written in the form (9.4.4a), and (9.3.18) shows that the update for the error vector could be written in the form (9.4.4b). Additionally, both sides of (9.4.4) can be premultiplied by any general vector z in $\{U, v\}$ to obtain the update recursion for the generalized inner product $\langle z, P_{Uv}^{\perp} y \rangle$. This operation provides

$$\langle z, P_{Uv} y \rangle = \langle z, P_U y \rangle + \langle z, P_U^{\perp} v \rangle \langle P_U^{\perp} v, P_U^{\perp} v \rangle^{-1} \langle v, P_U^{\perp} y \rangle, \quad (9.4.5a)$$

$$\langle z, P_{Uv}^{\perp} y \rangle = \langle z, P_U^{\perp} y \rangle - \langle z, P_U^{\perp} v \rangle \langle P_U^{\perp} v, P_U^{\perp} v \rangle^{-1} \langle v, P_U^{\perp} y \rangle. \quad (9.4.5b)$$

By a judicious selection of U, v, z, and y in (9.4.3)–(9.4.5), all time and order updates needed for both the LSL and transversal least squares filters may be derived. Each will be useful depending on whether a matrix, vector, or scalar least squares update is needed. Keep in mind that (9.4.3)–(9.4.5) were derived for the general least squares problem and are therefore applicable to both lattice and transversal forms. One type of update that the lattice and transversal filter have in common is the need to time update certain parameters. Since all of these will be based on the projection matrix, Section 9.5 derives a method for time updating the projection matrix.

9.5 Projection Matrix Time Update

The time update for the projection matrix is easily deduced from a simple example. Consider Figure 9.5, illustrating the geometrical relations for the least squares problem of predicting $d(n)$ using one basis vector $x(n)$. The time update results for this case are very intuitive geometrically, and may easily be extended to the general case of m basis vectors. For example purposes, let $\{d(n)\} = \{3, 1, 4, \ldots\}$ and $\{x(n)\} = \{4, 3, 2, \ldots\}$. Figure 9.5 illustrates the relations between the vectors $x(2)$ and $d(2)$, the vectors $x(3)$ and $d(3)$, and the unit vector $\pi(3) = [0, 0, 1]^T$. The vector $\pi(3)$ is the three-component case of

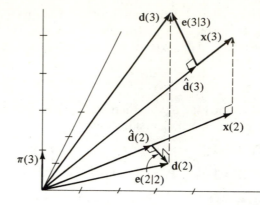

Figure 9.5   Geometrical construction of $\hat{\mathbf{d}}(2)$, $\hat{\mathbf{d}}(3)$ from $\mathbf{x}(2)$, $\mathbf{x}(3)$. $\{d(n)\} = \{3, 1, 4, \ldots\}$, $\{x(n)\} = \{4, 3, 2, \ldots\}$.

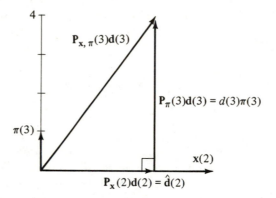

Figure 9.6   Projections of vectors from Figure 9.5 onto the $\{\mathbf{x}(3), \pi(3)\}$ subspace.

the n-component unit vector

$$\pi(n) = [0, 0, \ldots, 0, 1]^T. \tag{9.5.1}$$

Note that the vector $\pi(n)$ has unit length and points along the nth component axis. Therefore, the vector $\pi(n)$ is guaranteed to have a non-zero projection along the coordinate axis containing the current time n component of any vector. For this reason, $\pi(n)$, will be called the unit time vector, and its main use will be in providing a subspace for decomposing vectors into their "past" and "current" components. For the present example, this subspace will be $\{\mathbf{x}(3), \pi(3)\}$; that is, the subspace spanned by the current data vector and the unit time vector. The previous prediction $\hat{\mathbf{d}}(2)$ lies in this subspace, as must the updated prediction $\hat{\mathbf{d}}(3)$.

Now consider Figure 9.6, which shows the projections of various vectors from Figure 9.5 onto the $\{\mathbf{x}(3), \pi(3)\}$ subspace. It is easy to see that there is a right triangle formed by the previous prediction $\hat{\mathbf{d}}(2)$, the projection of $\mathbf{d}(3)$

onto $\pi(3)$, and the projection of $\mathbf{d}(3)$ onto $\{\mathbf{x}(3), \pi(3)\}$. This provides the equality

$$\mathbf{P}_{\mathbf{x},\pi}(3)\mathbf{d}(3) = \hat{\mathbf{d}}(2) + \mathbf{P}_{\pi}(3)\mathbf{d}(3). \tag{9.5.2}$$

Since $\hat{\mathbf{d}}(2)$ may be written in terms of the projection matrix $\mathbf{P}_{\mathbf{x}}(2)$, then (9.5.2) becomes

$$\mathbf{P}_{\mathbf{x},\pi}(3)\mathbf{d}(3) = \begin{bmatrix} \mathbf{P}_{\mathbf{x}}(2) & \mathbf{0}_2 \\ \mathbf{0}_2^T & 0 \end{bmatrix} \begin{bmatrix} \mathbf{d}(2) \\ d(3) \end{bmatrix} + \begin{bmatrix} \mathbf{0}_{2\times 2} & \mathbf{0}_2 \\ \mathbf{0}_2^T & 1 \end{bmatrix} \begin{bmatrix} \mathbf{d}(2) \\ d(3) \end{bmatrix}. \tag{9.5.3}$$

Since the matrices in (9.5.3) all operate on the same vector $\mathbf{d}(3)$, there is thus an equality between the matrices. Moreover, the equality holds for any general time n, giving

$$\mathbf{P}_{\mathbf{x},\pi}(n) = \begin{bmatrix} \mathbf{P}_{\mathbf{x}}(n-1) & 0 \\ 0 & 1 \end{bmatrix}. \tag{9.5.4}$$

The extension to a general m dimension data subspace $\{\mathbf{U}(n)\}$ then follows immediately:

$$\mathbf{P}_{\mathbf{U},\pi}(n) = \begin{bmatrix} \mathbf{P}_{\mathbf{U}}(n-1) & 0 \\ 0 & 1 \end{bmatrix}. \tag{9.5.5}$$

Equation (9.5.5) gives the desired result. Appending the $\pi(n)$ vector to the current data matrix provides a method for decomposing a projection matrix into its "past" and "current" components. This will be very valuable in the time update equations for the fast lattice and fast transversal filters.

Angle parameter

Forming projections onto subspaces containing the unit vector $\pi(n)$ also provides a useful geometrical interpretation of another important least squares parameter. Consider the orthogonal projection matrix update of (9.4.3b) letting $\mathbf{v} = \pi(n)$, and $\mathbf{U} = \mathbf{U}(n)$, a general data matrix at time n. It is easy to show using (9.5.5) that

$$\begin{bmatrix} \mathbf{P}_{\mathbf{U}}^{\perp}(n-1) & \mathbf{0}_{n-1} \\ \mathbf{0}_{n-1}^T & 0 \end{bmatrix} = \mathbf{P}_{\mathbf{U}}^{\perp}(n) - \frac{\mathbf{P}_{\mathbf{U}}^{\perp}(n)\mathbf{P}_{\pi}(n)\mathbf{P}_{\mathbf{U}}^{\perp}(n)}{\gamma_{\mathbf{U}}(n)}, \tag{9.5.6}$$

where $\gamma_{\mathbf{U}}(n)$ is the scalar angle parameter given by

$$\gamma_{\mathbf{U}}(n) = \langle \pi(n), \mathbf{P}_{\mathbf{U}}^{\perp}(n)\pi(n) \rangle. \tag{9.5.7}$$

To see the geometrical interpretation of $\gamma_{\mathbf{U}}(n)$, reconsider the previous example in which $\mathbf{U}(n) = \mathbf{x}(3)$. In this case, (9.5.7) becomes

$$\gamma_{\mathbf{x}}(3) = \langle \pi(3), \mathbf{P}_{\mathbf{x}}^{\perp}(3)\pi(3) \rangle \tag{9.5.8}$$

The parameters in (9.5.8) provide a very useful geometrical interpretation of $\gamma_{\mathbf{x}}(3)$, as shown in Figure 9.7. Let $\hat{\mathbf{i}}$ and $\hat{\mathbf{j}}$ be the two orthogonal unit vectors

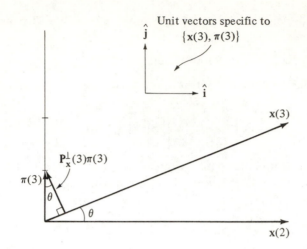

Figure 9.7   Angular relations among the least squares parameters in the $\{\mathbf{x}(3), \pi(3)\}$ subspace.

in the subspace $\{\mathbf{x}(3), \pi(3)\}$. In Figure 9.7, the vector $\hat{\mathbf{i}}$ points in the direction of $\mathbf{x}(2)$ and $\hat{\mathbf{j}}$ points in the direction of $\pi(3)$. Thus, $\mathbf{P}_{\mathbf{x}}^{\perp}(3)\pi(3)$ is the "projection of $\pi(3)$ orthogonal to $\mathbf{x}(3)$" and is a vector with $\hat{\mathbf{i}}, \hat{\mathbf{j}}$ components given by

$$\mathbf{P}_{\mathbf{x}}^{\perp}(3)\pi(3) = \cos\theta \begin{bmatrix} -\hat{\mathbf{i}}\sin\theta \\ \hat{\mathbf{j}}\cos\theta \end{bmatrix}, \qquad (9.5.9)$$

where θ is the angle between the basis vectors $\mathbf{x}(2)$ and $\mathbf{x}(3)$. By the same process, $\pi(3)$ may be represented by the vector

$$\pi(3) = \begin{bmatrix} 0 \\ \hat{\mathbf{j}} \end{bmatrix}, \qquad (9.5.10)$$

and therefore $\gamma_{\mathbf{x}}(3)$ from (9.5.8) becomes

$$\gamma_{\mathbf{x}}(3) = \cos^2\theta. \qquad (9.5.11)$$

Thus, the parameter $\gamma_{\mathbf{x}}(3)$ is directly related to the angular change between $\mathbf{x}(2)$, which is the data subspace at time $n = 2$, and $\mathbf{x}(3)$, which is the data subspace at time $n = 3$. In general, $\gamma_{\mathbf{U}}(n)$ for the m dimension case will have a similar interpretation of quantifying the angular change between the m dimension subspace $\{\mathbf{U}(n-1)\}$ and $\{\mathbf{U}(n)\}$. As the basis vectors of $\{\mathbf{U}(n-1)\}$ add their nth components and thus change to those of $\{\mathbf{U}(n)\}$, then the resulting subspace changes its orientation in some manner. In higher dimension spaces, this is harder to visualize than in (9.5.11), but the concept is similar. This angle parameter $\gamma_{\mathbf{U}}(n)$ will be very useful in the specific time update recursions for the lattice and transversal filters. The "U" subscript will change to reflect the specific data subspace being examined.

As another specific example of this angle parameter, consider the extension of (9.5.8) to the m dimension data subspace $\mathbf{X}_{0,m-1}(n)$ previously examined. In a manner similar to (9.5.8), the angle parameter $\gamma_m(n)$ would then be defined as

$$\gamma_m(n) = \langle \pi(n), \mathbf{P}_{0,m-1}^{\perp}(n)\pi(n)\rangle, \tag{9.5.12}$$

where the "m" reflects the m dimension data subspace. Thus, $\gamma_m(n)$ quantifies the angular change between the successive data subspaces $\{\mathbf{X}_{0,m-1}(n-1)\}$ and $\{\mathbf{X}_{0,m-1}(n)\}$. For these m dimension subspaces, the angular change is actually the angular change between the unit vectors normal to the subspaces.

Another relation that will be very useful in work to come is obtained by directly substituting $n - 1$ for n in (9.5.12). After simplification, this produces

$$\gamma_m(n-1) = \langle \pi(n), \mathbf{P}_{1,m}^{\perp}(n)\pi(n)\rangle. \tag{9.5.13}$$

The relation between (9.5.12) and (9.5.13) will be very valuable in the time updates for both lattice and transversal least squares parameters.

PROBLEMS

1. Using the prewindowed error as defined by (9.3.2), derive the expression given in (9.3.10) for the optimal predictor, $\mathbf{w}_m(n)$, by minimizing the prewindowed error vector norm.

2. This problem displays some of the geometrical concepts associated with least squares filtering.
   (a) Suppose the time series $x(n) = 2, 3, 4, 6, -1, \ldots$ is used as the input to the least squares filter to make a prediction of another time series $d(n)$. Draw the basis vectors used for the linear prediction for the case of $m = 2$ and $m = 3$.
   (b) Suppose now a different time series $x(n) = 4, 0, 0, 0, \ldots$ is used as the input to the least squares filter. Draw the basis vectors corresponding to the case $m = 2$, $n = 3$.

3. This problem expands upon the results of P2.
   (a) Find the projection matrix $\mathbf{P}_{0,1}(3)$ for the time series in P2(a) above.
   (b) Find the projection matrix $\mathbf{P}_{0,1}(3)$ for the time series in P2(b) above.

4. This problem provides experience in actually computing the optimal least squares filter coefficients for a specific set of data.
   (a) Using the projection matrix derived in P3(a), find the least squares filter, $\mathbf{w}_2(3)$, and the least squares prediction for the "desired" vector $\mathbf{d}(n) = [4, 6, 8]^T$.
   (b) Using the projection matrix derived in P3(b), find the least squares filter, $\mathbf{w}_2(3)$, and the least squares prediction for the desired vector $\mathbf{d}(n) = [4, 6, 8]^T$.
   (c) Explain why the least squares filters in (a) and (b) above are different, even though the desired vector is the same in both cases.

5. Adding basis vectors to the subspace onto which the least squares estimate is projected will decrease the error vector norm until enough basis vectors have been incorporated to span the space of the desired vector. This problem illustrates this decrease in the error norm.
   (a) Using only $m = 1$ basis vector, given by $\mathbf{x}(3) = [5, 3, 1]^T$, find the norm of the error vector, $\mathbf{e}_1(3|3)$, resulting from a prediction of the desired vector $\mathbf{d}(3) = [3, 3, 3]^T$.

(b) Now add the second basis vector, $z^{-1}\mathbf{x}(3) = [0, 5, 3]^T$, and find the norm of $\mathbf{e}_2(3|3)$, which results from making a prediction of $\mathbf{d}(3)$ from part (a).

(c) Verify that the error vectors in parts (a) and (b) are orthogonal to the respective subspaces containing the predictions.

6. Show that the cumulative squared error criterion of (8.2.4) can be written in the form

$$\varepsilon(n) = \langle \mathbf{A}_{nn}\mathbf{e}(n|n), \mathbf{A}_{nn}\mathbf{e}(n|n) \rangle,$$

where $\mathbf{A}_{nn}$ is an $n \times n$ diagonal matrix. What are the elements of $\mathbf{A}_{nn}$?

7. This problem explores the block-processing approach to computing the least squares filter and the least squares prediction. Let K points of data and desired signal be acquired according to the technique of (9.3.10) for $n = K$. Assume that p^3 arithmetic operations (a.o.) are needed to invert a $p \times p$ matrix. Given this, how many a.o would be needed to compute $\mathbf{w}_m(K)$? How many to compute $\hat{\mathbf{d}}(K|K)$? What is the limitation of this approach as $K \to \infty$?

8. Consider the square, partitioned matrix $\mathbf{H}_{m+1,m+1}$ and its partitioned inverse $\mathbf{H}_{m+1,m+1}^{-1}$ as given below:

$$\mathbf{H}_{m+1,m+1} = \begin{bmatrix} \mathbf{A}_{mm} & \mathbf{b}_m \\ \mathbf{c}_m^T & d \end{bmatrix}, \qquad \mathbf{H}_{m+1,m+1}^{-1} = \begin{bmatrix} \mathbf{Q}_{mm} & \mathbf{r}_m \\ \mathbf{s}_m^T & t \end{bmatrix}.$$

Assume that d is non-zero and that $\mathbf{A}_{mm}^{-1}$ exists and is known. Find the elements $\mathbf{Q}_{mm}$, $\mathbf{r}_m$, $\mathbf{s}_m^T$, and t in terms of the known elements $\mathbf{A}_{mm}^{-1}$, $\mathbf{b}_m$, $\mathbf{c}_m^T$, and d.

9. Show that the following relations are true:

(a) $\mathbf{X}_{1,k}(n) = \begin{bmatrix} 0_k^T \\ \mathbf{X}_{0,k-1}(n-1) \end{bmatrix}$,   (b) $\mathbf{P}_{1,k}(n) = \begin{bmatrix} 0 & 0_{n-1}^T \\ 0_{n-1} & \mathbf{P}_{0,k-1}(n-1) \end{bmatrix}$,

where 0_q is the "q-component vector of all zeros."

10. Find an expression for the following matrix inverse

$$\langle \mathbf{X}_{1,p}(n), \mathbf{X}_{1,p}(n) \rangle^{-1}$$

in terms of the matrix inverse

$$\langle \mathbf{X}_{1,p}(n-1), \mathbf{X}_{1,p}(n-1) \rangle^{-1}.$$

11. Suppose the two basis vectors

$$\mathbf{u}_1 = [-1, 2, -4, 3, 1]^T, \qquad \mathbf{u}_2 = [5, 6, 2, -2, -1]^T$$

span the vector space $\{\mathbf{U}\} = \{\mathbf{u}_1, \mathbf{u}_2\}$. Prove the vector

$$\mathbf{v} = [-31, -18, -34, 28, 11]^T$$

either does or does not lie in $\{\mathbf{U}\}$.

REFERENCES

1. G.H. Golub and C.F. Van Loan, *Matrix Computations*, Johns Hopkins University Press, Baltimore, MD, 1983.

2. P. Lancaster, *Theory of Matrices*, Academic Press, New York, 1969.
3. R. Bellman, *Introduction to Matrix Analysis*, McGraw-Hill, New York, 1960.
4. D.G. Luenberger, *Optimization by Vector Space Methods*, John Wiley & Sons, New York, 1969.
5. P. Halmos, *Finite Dimensional Vector Spaces*, Van Nostrand, Princeton, NJ, 1958.
6. P. DeRusso, R.B. Roy, and C.M. Close, *State Variables for Engineers*, John Wiley & Sons, New York, 1974.
7. G. Stewart, *Introduction to Matrix Computations*, Academic Press, New York, 1969.
8. C.L. Lawson and R.J. Hanson, *Solving Least Squares Problems*, Prentice-Hall, Englewood Cliffs, NJ, 1974.
9. D.C. Youla, "Generalized Image Restoration by the Method of Alternating Orthogonal Projections," *IEEE Trans. on Circuits and Sys.*, vol. CAS-25, no. 9, pp. 694–702 September 1978.
10. A.P. Naylor and J.N. Sell, *Linear Operator Theory in Science and Engineering*, Springer-Verlag, New York, 1976.
11. T.C. Hsia, *Systems Identification: Least Squares Methods*, Lexington Press, Lexington, MA, 1977.
12. B. Noble, *Applied Linear Algebra*, Prentice-Hall, Englewood Cliffs, NJ, 1969.
13. A. Albert, *Regression and the Moore-Penrose Pseudoinverse*, Academic Press, New York, 1972.

CHAPTER 10

The Least Squares Lattice Algorithm

10.1 Introduction

This chapter will examine the derivation of the least squares lattice (LSL) adaptive filter through the use of vector space concepts. As seen from Chapters 3 and 7, the lattice form is a specific implementation for the Nth order linear prediction filter. It is order recursive in nature, meaning that all lower order prediction filters are also computed in the process of calculating the Nth order filter. After the error minimization problem has been formulated properly in the appropriate vector space, the resulting derivation for the LSL equations is straightforward. For simplicity and to introduce the application of vector space concepts, this chapter investigates the use of the lattice as a linear predictor.

The introduction of the LSL using vector space concepts is fairly recent. Lee and Morf [1, 2] provided a basic structure for the algorithm, which was then expanded later [3]. Additionally, a comparable LSL structure can also be derived by matrix algebraic techniques [4, 5]. Satorius and Pack [6] applied the LSL to the linear channel equalization problem, and more recently Ling and Proakis [7, 8] applied the LSL to the decision-feedback equalizer problem. Other applications include signal tracking by Hodgkiss and Presley [9], spectrum analysis by Reddy, et al. [10], and speech pitch estimation by Lee and Morf [11]. Finally, two survey works on the derivations and applications of the LSL are those by Friedlander [5] and Turner [12].

This chapter is organized as follows. First, Section 10.2 uses the vector space relations of Chapter 9 to define the forward and backward prediction error filters. Section 10.3 then shows that the lattice structure is a natural result of minimizing the least squares (LS) forward prediction error criterion.

Derivations of the necessary time and order updates required for the resulting LS lattice are then presented in Section 10.4. Finally, Section 10.5 applies the LS lattice to a simple linear prediction problem to show the increase in performance compared to the least mean squares (LMS) algorithm.

10.2 Forward and Backward Prediction Filters

Previous work from Chapters 3 and 7 has shown that one implementation of the Nth order linear prediction filter is the lattice structure. In the lattice, only the reflection coefficients are required to be updated as a function of time and order. Therefore, one viable approach would be to assume a lattice implementation of the LS linear predictor, and then derive the required update recursions such that the orthogonality conditions were maintained. However, the lattice structure is also a straightforward consequence of simply computing the $(m + 1)$st order forward linear prediction filter based upon a knowledge of the mth order linear prediction filter. This will be shown in detail in Section 10.3. It should be noted that requiring a lattice implementation for the linear predictor is no more restrictive than requiring a transversal implementation since both are assumptions of a specific filter implementation.

The forward and backward filters of this section are similar to the forward and backward prediction filters of Chapter 3. A block diagram of a single stage of an LS lattice is shown in Figure 10.1(a), and a cascade of N stages to form the Nth order linear prediction filter is shown in Figure 10.1(b). The main difference between the LS lattice and the lattice of Figure 3.1 is that, in general, the LS reflection coefficients $k_m^f(n)$ and $k_m^b(n)$ are not equal. This is admissible since no stationarity requirement for the data has been assumed in generating the LS error criterion. For each time iteration n, the lattice implementation of the Nth order LS prediction filter provides forward prediction errors, $e_m^f(n)$, and backward prediction errors, $e_m^b(n)$, for all orders $1 \leq m \leq N$.

Forward prediction error (FPE) filter

Using the vector space interpretation of Chapter 9, the LS filters are seen to predict entire n component signal vectors instead of simple scalars. Therefore, the mth order forward prediction error (FPE) filter computes a prediction, $\hat{\mathbf{x}}(n)$, of the present data vector $\mathbf{x}(n)$ based upon the m basis vectors of the subspace $\{\mathbf{X}_{1,m}(n)\}$. Note that the basis vectors of $\{\mathbf{X}_{1,m}(n)\}$ are all delayed vectors, and thus $\mathbf{x}(n)$ does not lie in $\{\mathbf{X}_{1,m}(n)\}$. Hence, no information about the present sample $x(n)$ is contained in $\{\mathbf{X}_{1,m}(n)\}$. In vector form

$$\mathbf{e}_m^f(n) = \mathbf{x}(n) - \hat{\mathbf{x}}(n). \tag{10.2.1}$$

The results of Chapter 9 immediately require the optimal LS prediction of $\mathbf{x}(n)$ to be simply the projection of $\mathbf{x}(n)$ onto the subspace $\{\mathbf{X}_{1,m}(n)\}$, or

$$\hat{\mathbf{x}}(n) = \mathbf{P}_{1,m}(n)\mathbf{x}(n), \tag{10.2.2}$$

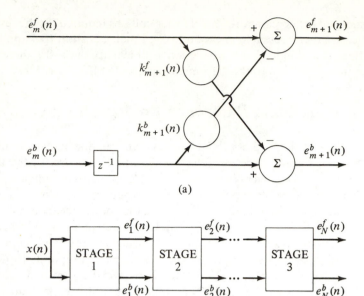

(a)

(b)

Figure 10.1   (a) Single stage of the least squares lattice filter.   (b) Cascade of stages to form Nth order linear prediction filter.

where $\mathbf{P}_{1,m}(n)$ is the projection matrix

$$\mathbf{P}_{1,m}(n) = \mathbf{X}_{1,m}(n)\langle \mathbf{X}_{1,m}(n), \mathbf{X}_{1,m}(n)\rangle^{-1}\mathbf{X}_{1,m}^{T}(n). \tag{10.2.3}$$

The mth order FPE vector is then given simply by the component of $\mathbf{x}(n)$ orthogonal to $\{\mathbf{X}_{1,m}(n)\}$, or

$$\mathbf{e}_m^f(n) = \mathbf{P}_{1,m}^{\perp}(n)\mathbf{x}(n), \tag{10.2.4}$$

where

$$\mathbf{P}_{1,m}^{\perp}(n) = \mathbf{I}_{nn} - \mathbf{P}_{1,m}(n)$$

is the orthogonal projection matrix. The current time FPE scalar sample is easily derived as

$$e_m^f(n) = \langle \pi(n), \mathbf{e}_m^f(n)\rangle = \langle \pi(n), \mathbf{P}_{1,m}^{\perp}(n)\mathbf{x}(n)\rangle. \tag{10.2.5}$$

The benefit of writing $e_m^f(n)$, as in (10.2.5), is that the FPE is now in the form required by (9.4.5b) for the LS update of a scalar. This will be very important in the order updates for the LS lattice.

Backward prediction error (BPE) filter

In a similar fashion, the backward prediction error (BPE) filter may now be developed from the standpoint of a projection onto a slightly different subspace. Recall from the previous work in Chapters 3 and 7 that the backward filter computes $\hat{x}_b(n - m)$, a prediction of the scalar sample $x(n - m)$, based upon the m most recent samples occurring after $x(n - m)$. This led to the BPE:

$$e_m^b(n) = x(n - m) - \hat{x}_b(n - m). \tag{10.2.6}$$

Considering the vector of all backward error from time 1 to time n, the vector relation analogous to (10.2.6) becomes

$$\begin{aligned}
\mathbf{e}_m^b(n) &= z^{-m}\mathbf{x}(n) - \hat{\mathbf{x}}_b(n - m) \\
&= [e_m^b(1), e_m^b(2), \dots, e_m^b(n)]^T.
\end{aligned} \tag{10.2.7}$$

The vector space approach of Chapter 9 may be used to find the form of $\hat{\mathbf{x}}_b(n - m)$ in a very simple manner. To compute the backward prediction vector with a linear filter, use a linear combination of the m data vectors, which immediately follow $z^{-m}\mathbf{x}(n)$ in time. These vectors are therefore the m basis vectors that define the m dimension subspace $\{\mathbf{X}_{0,m-1}(n)\}$ where

$$\mathbf{X}_{0,m-1}(n) = [z^0\mathbf{x}(n)z^{-1}\mathbf{x}(n)\cdots z^{-m+1}\mathbf{x}(n)]. \tag{10.2.8}$$

Using the previous development, the backward prediction is then given simply by:

$$\hat{\mathbf{x}}_b(n - m) = \mathbf{P}_{0,m-1}(n)z^{-m}\mathbf{x}(n), \tag{10.2.9}$$

where $\mathbf{P}_{0,m-1}(n)$ is the projection matrix

$$\mathbf{P}_{0,m-1}(n) = \mathbf{X}_{0,m-1}(n)\langle\mathbf{X}_{0,m-1}(n), \mathbf{X}_{0,m-1}(n)\rangle^{-1}\mathbf{X}_{0,m-1}^T(n). \tag{10.2.10}$$

Note that the structure of the computation in (10.2.10) is identical to that in (10.2.3). Only the basis vectors of the subspace have changed. In general, a projection matrix may be created for any desired subspace by simply redefining the basis vectors of the subspace.

Lastly, the backward error vector is the part of $z^{-m}\mathbf{x}(n)$ that is not predictable using $\{\mathbf{X}_{0,m-1}(n)\}$, and is therefore the projection of $z^{-m}\mathbf{x}(n)$ orthogonal to $\{\mathbf{X}_{0,m-1}(n)\}$. This is easily seen from (10.2.7) and (10.2.9) to be:

$$\mathbf{e}_m^b(n) = [\mathbf{I}_{nn} - \mathbf{P}_{0,m-1}(n)]z^{-m}\mathbf{x}(n) = \mathbf{P}_{0,m-1}^{\perp}(n)z^{-m}\mathbf{x}(n), \tag{10.2.11}$$

where $\mathbf{P}_{0,m-1}^{\perp}(n)$ is the orthogonal projection matrix. Similar to (10.2.6), the current scalar BPE is given by

$$e_m^b(n) = \langle\pi(n), \mathbf{e}_m^b(n)\rangle = \langle\pi(n), \mathbf{P}_{0,m-1}^{\perp}(n)z^{-m}\mathbf{x}(n)\rangle. \tag{10.2.12}$$

FPE and BPE residuals

The squared norms of the FPE and BPE vectors will be very useful in work to come. These are commonly called the BPE and FPE residuals in the literature, and are a measure of the FPE and BPE energy at the mth stage of the lattice. The FPE residual is a scalar and is defined as

$$\varepsilon_m^f(n) = \langle \mathbf{e}_m^f(n), \mathbf{e}_m^f(n) \rangle, \tag{10.2.13}$$

while the BPE residual is given by

$$\varepsilon_m^b(n) = \langle \mathbf{e}_m^b(n), \mathbf{e}_m^b(n) \rangle. \tag{10.2.14}$$

Using the shift properties of the z^{-1} operator then produces two additional useful relations for the residuals at the previous time sample:

$$\varepsilon_m^f(n-1) = \langle z^{-1}\mathbf{e}_m^f(n), z^{-1}\mathbf{e}_m^f(n) \rangle, \tag{10.2.15}$$

$$\varepsilon_m^b(n-1) = \langle z^{-1}\mathbf{e}_m^b(n), z^{-1}\mathbf{e}_m^b(n) \rangle, \tag{10.2.16}$$

10.3 The LS Lattice Structure

The lattice structure results as a consequence of computing the $(m+1)$st order LS forward linear predictor based upon a knowledge of the mth order LS linear predictor. However, no direct specification of filter form has been made at this time, and there is nothing as yet to suggest the lattice form. However, since (10.2.6) holds for any order, it holds for order $m+1$ and thus

$$e_{m+1}^f(n) = \langle \pi(n), \mathbf{P}_{0,m+1}^{\perp}(n)\mathbf{x}(n) \rangle. \tag{10.3.1}$$

The form (10.3.1) defines $e_{m+1}^f(n)$ as a scalar resulting from an LS minimization problem. Therefore, (9.4.5b) immediately applies, and it is only necessary to determine the vectors $\mathbf{v}$, $\mathbf{y}$, and $\mathbf{z}$ in (9.4.5b). Since the $(m+1)$st order linear predictor now accesses the scalar sample $x(n-m-1)$ one sample earlier in time, the vector space analogy is that the data vector $z^{-m-1}\mathbf{x}(n)$ is now included in the data subspace. Thus, for the $(m+1)$st order linear predictor, the data subspace is $\{\mathbf{X}_{1,m+1}(n)\}$, which is obtained by appending the vector $\mathbf{v} = z^{-m-1}\mathbf{x}(n)$ in (9.4.5b) to the old data matrix $\mathbf{U} = \mathbf{X}_{1,m}(n)$. The $\mathbf{y}$ vector in (9.4.5b) is also seen to be $\mathbf{x}(n)$.

Another form that appears from making these associations in (9.4.5b) is

$$\mathbf{P}_{\mathbf{U}}^{\perp}\mathbf{v} = \mathbf{P}_{1,m}^{\perp}(n)z^{-m-1}\mathbf{x}(n) = z^{-1}\mathbf{e}_m^b(n), \tag{10.3.2}$$

where $z^{-1}\mathbf{e}_m^b(n)$ is the delayed (down-shifted) backward error vector at time n. The right-hand side of (10.3.2) may be derived by expanding the left-hand side and using the definition of the BPE vector. The components of $z^{-1}\mathbf{e}_m^b(n)$ are given by

$$z^{-1}\mathbf{e}_m^b(n) = [0, e_m^b(1), \ldots, e_m^b(n-1)]^T. \tag{10.3.3}$$

Therefore, making the substitutions in (9.4.5b) gives the intermediate result

$$e_{m+1}^f(n) = e_m^f(n) - \frac{\langle \pi(n), z^{-1}\mathbf{e}_m^b(n) \rangle \langle z^{-1}\mathbf{e}_m^b(n), \mathbf{x}(n) \rangle}{\varepsilon_m^b(n-1)}, \tag{10.3.4}$$

where (10.2.16) has also been used. Then using (10.3.3), the inner product in (10.3.4) containing $\pi(n)$ produces

$$\langle \pi(n), z^{-1} \mathbf{e}_m^b(n) \rangle = e_m^b(n-1), \tag{10.3.5}$$

which is the scalar BPE at the previous time iteration. The vector $\mathbf{x}(n)$ in the other inner product of (10.3.4) may be expanded using (10.2.1) and (10.2.2) to give

$$\langle z^{-1} \mathbf{e}_m^b(n), \mathbf{e}_m^f(n) + \mathbf{P}_{1,m}(m)\mathbf{x}(n) \rangle = \langle z^{-1} \mathbf{e}_m^b(n), \mathbf{e}_m^f(n) \rangle, \tag{10.3.6}$$

where the relations (9.2.8) and (9.3.24) have also been used.

The inner product (10.3.6) is very important in the lattice LS predictor and is called the partial correlation (PARCOR) coefficient, $\Delta_{m+1}(n)$, between forward and backward prediction errors:

$$\Delta_{m+1}(n) = \langle z^{-1} \mathbf{e}_m^b(n), \mathbf{e}_m^f(n) \rangle. \tag{10.3.7}$$

Therefore, using (10.3.7) and (10.3.5) in (10.3.4) gives the following order update equation for the $(m+1)$st order FPE:

$$e_{m+1}^f(n) = e_m^f(n) - k_{m+1}^b(n)e_m^b(n-1), \tag{10.3.8}$$

where the $(m+1)$st order backward reflection coefficient, $k_{m+1}^b(n)$, is defined by

$$k_{m+1}^b(n) = \frac{\Delta_{m+1}(n)}{\varepsilon_m^b(n-1)}. \tag{10.3.9}$$

From (10.3.8), it is seen that the expression for computing the FPE at the $(m+1)$st order requires the "backward" prediction error $e_m^b(n-1)$. This backward prediction error appears as a straightforward consequence of applying the LS update (9.4.5b) for an inner product, and is due solely to the minimization of the forward LS error criterion. However, also performing a recursive computation of the BPE leads to the lattice form in Figure 10.1 for the resulting linear predictor. Therefore, to complete the computation of the updated FPE in (10.3.8) requires first computing the order update for the BPE.

This is done by making the associations $\mathbf{U} = \mathbf{X}_{1,m}(n)$, $\mathbf{v} = \mathbf{x}(n)$, and $\mathbf{y} = z^{-m-1}\mathbf{x}(n)$ in (9.4.5b) and completing a derivation similar to the preceding for the FPE. One point to recognize is that the subspaces $\{\mathbf{U}, \mathbf{v}\}$ and $\{\mathbf{v}, \mathbf{U}\}$ are equivalent; that is,

$$\{\mathbf{X}_{1,m}(n), \mathbf{x}(n)\} = \{\mathbf{x}(n), \mathbf{X}_{1,m}(n)\} = \{\mathbf{X}_{0,m}(n)\}. \tag{10.3.10}$$

Using this result, $\{\mathbf{U}, \mathbf{v}\} = \{\mathbf{X}_{0,m}(n)\}$ in (9.4.5b), and the derivation of backward prediction error order update is straightforward. The result is given by

$$e_{m+1}^b(n) = e_m^b(n-1) - k_{m+1}^f(n)e_m^f(n), \tag{10.3.11}$$

where similar to (10.3.9) $k_{m+1}^f(n)$ is the forward reflection coefficient,

$$k_{m+1}^f(n) = \frac{\Delta_{m+1}(n)}{\varepsilon_m^f(n)}, \tag{10.3.12}$$

and $\varepsilon_m^f(n)$ is the FPE residual from (10.2.13).

Equations (10.3.8) and (10.3.11) define the lattice structure of Figure 10.1.

To implement this structure in an order-recursive and time-recursive manner only requires the reflection coefficients to be updated. In turn, from (10.3.11) and (10.3.12), this means that only three parameters need to be computed for each new order or each new time iteration: $\varepsilon_m^f(n)$, $\varepsilon_m^b(n)$, and $\Delta_{m+1}(n)$. If these three parameters can be updated in either an order-recursive or a time-recursive manner, then the LS lattice can be implemented. Therefore, the remaining derivation of the LS lattice focuses on updating these parameters.

10.4 Lattice Order and Time Updates

Order update equations for the FPE and BPE residuals are straightforward and follow from making the appropriate associations in (9.4.5b). Using (10.2.4), (10.2.13), and the properties of the projection matrix, it is simple to show

$$\varepsilon_m^f(n) = \langle \mathbf{x}(n), \mathbf{P}_{1,m}^\perp(n)\mathbf{x}(n) \rangle. \tag{10.4.1}$$

A similar processing using (10.2.11) and (10.2.14) gives a relation for the BPE residual

$$\varepsilon_m^b(n) = \langle z^{-m}\mathbf{x}(n), \mathbf{P}_{0,m-1}^\perp(n)z^{-m}\mathbf{x}(n) \rangle. \tag{10.4.2}$$

Therefore, the order update for $\varepsilon_{m+1}^f(n)$ is given from (9.4.5b) by using $\mathbf{z} = \mathbf{y} = \mathbf{x}(n)$, $\mathbf{U} = \mathbf{X}_{1,m}(n)$, and $\mathbf{v} = z^{-m+1}\mathbf{x}(n)$. After simplification, this produces

$$\varepsilon_{m+1}^f(n) = \varepsilon_m^f(n) - \frac{\Delta_{m+1}^2(n)}{\varepsilon_m^b(n-1)}. \tag{10.4.3}$$

A similar approach using the associations $\mathbf{z} = \mathbf{y} = z^{-m-1}\mathbf{x}(n)$, $\mathbf{U} = \mathbf{X}_{1,m}(n)$, and $\mathbf{v} = \mathbf{x}(n)$ then produces the order update for the BPE residual

$$\varepsilon_{m+1}^b(n) = \varepsilon_m^b(n-1) - \frac{\Delta_{m+1}^2(n)}{\varepsilon_m^f(n)}. \tag{10.4.4}$$

The relation in (10.3.6) is also useful in deriving the form (10.4.4).

Therefore, the two order updates (10.4.3) and (10.4.4) are now available. However, an attempt to extend this order update process to derive $\Delta_{m+2}(n)$ from $\Delta_{m+1}(n)$ produces a form $\mathbf{P}_{1,m}^\perp(n)z^{-m-2}\mathbf{x}(n)$ in the resulting associations. Unfortunately, this form is not any of the previously defined LS parameters. However, if a time update relation for $\Delta_{m+1}(n)$ is found instead, then a viable recursion for $\Delta_{m+1}(n)$ in terms of $\Delta_{m+1}(n-1)$ does result.

This is obtained by making the associations $\mathbf{v} = \pi(n)$, $\mathbf{U} = \mathbf{X}_{1,m}(n)$, $\mathbf{z} = \mathbf{x}(n)$, and $\mathbf{y} = z^{-m-1}\mathbf{x}(n)$ in (9.4.5b). After simplification, this produces

$$\Delta_{m+1}(n) = \Delta_{m+1}(n-1) + \frac{e_m^f(n)e_m^b(n-1)}{\gamma_m(n-1)}, \tag{10.4.5}$$

where $\gamma_m(n-1)$ is the angle parameter defined from (9.5.13),

$$\gamma_m(n-1) = \langle \pi(n), \mathbf{P}_{1,m}^\perp(n)\pi(n) \rangle. \tag{10.4.6}$$

In (10.4.5), all the parameters are now available and $\Delta_{m+1}(n)$ may be computed. However, for the next stage iteration, an order update for $\gamma_{m+1}(n-1)$ must be derived. This is the final update needed in the LS lattice recursion. Using $\mathbf{z} = \mathbf{y} = \pi(n)$, $\mathbf{U} = \mathbf{X}_{1,m}(n)$, and $\mathbf{v} = z^{-m-1}\mathbf{x}(n)$ in (9.4.5b), this gives

$$\gamma_{m+1}(n-1) = \gamma_m(n-1) - \frac{[e_m^b(n-1)]^2}{\varepsilon_m^b(n-1)}. \tag{10.4.7}$$

At this point, all the recursions needed to implement the LS lattice algorithm have been derived. For easy reference, these steps are listed in order in Table 10.1, together with initialization values. The parameter δ should be close to the steady-state squared prediction error, if this value is known. However, the value of δ is not critical, as will be shown in the following section.

Table 10.1 The Least Squares Lattice Algorithm

Initialize:

$$e_m^b(0) = \Delta_m(0) = 0, \tag{10.4.8}$$

$$\gamma_m(0) = 1, \tag{10.4.9}$$

$$\varepsilon_m^f(0) = \varepsilon_m^b(0) = \delta. \tag{10.4.10}$$

For $n = 1$ to n *final* do:

$$e_0^b(n) = e_0^f(n) = x(n), \tag{10.4.11}$$

$$\varepsilon_0^b(n) = \varepsilon_0^f(n) = \varepsilon_0^f(n-1) + x^2(n), \tag{10.4.12}$$

$$\gamma_0(n) = 1. \tag{10.4.13}$$

For $0 \le m \le N - 1$ do:

$$\Delta_{m+1}(n) = \Delta_{m+1}(n-1) + \frac{e_m^b(n-1)e_m^f(n)}{\gamma_m(n-1)}, \tag{10.4.14}$$

$$e_{m+1}^f(n) = e_m^f(n) - \frac{\Delta_{m+1}(n)e_m^b(n-1)}{\varepsilon_m^b(n-1)}, \tag{10.4.15}$$

$$e_{m+1}^b(n) = e_m^b(n-1) - \frac{\Delta_{m+1}(n)e_m^f(n)}{\varepsilon_m^f(n)}, \tag{10.4.16}$$

$$\varepsilon_{m+1}^f(n) = \varepsilon_m^f(n) - \frac{\Delta_{m+1}^2(n)}{\varepsilon_m^b(n-1)}, \tag{10.4.17}$$

$$\varepsilon_{m+1}^b(n) = \varepsilon_m^b(n-1) - \frac{\Delta_{m+1}^2(n)}{\varepsilon_m^f(n)}, \tag{10.4.18}$$

$$\gamma_{m+1}(n-1) = \gamma_m(n-1) - \frac{[e_m^b(n-1)]^2}{\varepsilon_m^b(n-1)}. \tag{10.4.19}$$

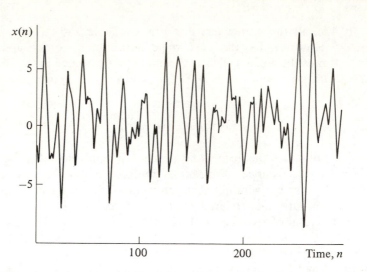

Figure 10.2   Second-order autoregressive random process $x(n)$: $a_1 = 1.558$, $a_2 = -0.81$, and $\sigma_v^2 = 1.0$.

10.5 Examples of LS Lattice Performance

This section presents some basic examples for using the LS lattice in a linear prediction application. Consider Figure 10.2, which shows a second-order autoregressive (AR) time signal created by passing an uncorrelated unity variance gaussian signal $v(n)$ through the linear system with poles at z_1, $z_2 = 0.9e^{j\pi/6}$, $0.9e^{-j\pi/6}$. It is easy to show that this leads to the signal model

$$x(n) = a_1 x(n-1) + a_2 x(n-2) + v(n), \qquad (10.5.1)$$

where the particular choice of poles gives $a_1 = 1.558$ and $a_2 = -0.81$.

It is desired to use the LS lattice in a linear prediction mode to estimate a_1 and a_2. This is illustrated in Figure 10.3, which displays the performance of the LS lattice and the LMS algorithm in the linear prediction mode. Since LMS implements a transversal filter, it provides an estimate of a_1 and a_2 directly. However, since the LS lattice does not implement a tapped delay line directly, it is necessary to compute the AR estimates as a function of the LS reflection coefficients. By expanding the lattice equations for $e_2^f(n)$, it is easy to show that the first and second AR coefficient estimates are given by

$$\hat{a}_1(n) = k_1^b(n) - k_1^f(n)k_2^b(n)$$
$$\hat{a}_2(n) = k_2^b(n), \qquad (10.5.2)$$

where $k_m^b(n)$ and $k_m^f(n)$ are the mth backward and forward reflection coefficients, given by (10.3.9) and (10.3.12), respectively.

These estimates are shown in Figure 10.3 using $\delta = 1.0$ for the LS lattice, corresponding to unity variance of the system excitation $v(n)$ in (10.5.1).

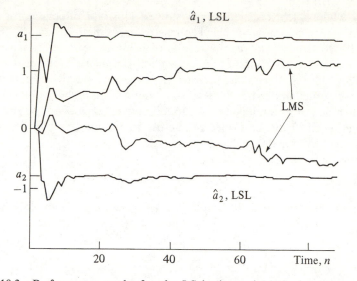

Figure 10.3   Performance results for the LS lattice and LMS algorithm as linear predictors. $a_1 = 1.558$, $a_2 = -0.81$.

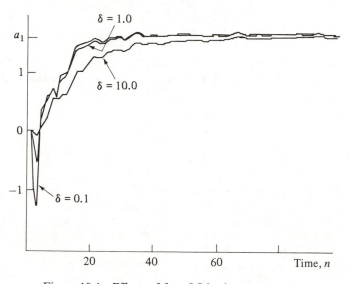

Figure 10.4   Effects of δ on LS lattice convergence.

This excitation is sometimes called the *innovations* of the AR process and is completely unpredictable. Therefore, in the steady state, no prediction method could expect to have average prediction error power less than the power of the excitation. Also shown in Figure 10.3 is the "best" of a series of runs using LMS to estimate a_1 and a_2. The gain value $\alpha = 0.01$ was used, which was the largest α value before divergence. The LS lattice is seen to converge extremely rapidly, whereas LMS is significantly slower. This type of behavior is typical of performance comparisons between least squares methods and LMS.

Figure 10.4 then displays the results of using three different δ values to estimate a_1 using the LS lattice linear predictor of the previous problem. The values $\delta = 0.1$ and $\delta = 1.0$ are seen to converge very rapidly to a_1, although they have some "undershoot" at the beginning. The higher values $\delta = 10.0$ mitigate much of the undershoot, although convergence is slowed somewhat. However, each of the LS lattice predictors in Figure 10.4 converges much faster than LMS for this problem.

PROBLEMS

1. Using the definition of the BPE (10.2.6) and a backward predictor of the form

$$\hat{x}_b(n - m) = \sum_{k=1}^{m} b_k x(n - m + k),$$

   derive the form given for the backward error vector given by (10.2.11).

2. If the jth order LS backward prediction error vector is given by

$$\mathbf{P}_{0, j-1}^{\perp}(n)z^{-j}\mathbf{x}(n) = \mathbf{e}_j^b(n),$$

   show that

$$\mathbf{P}_{1, j}^{\perp}(n)z^{-j-1}\mathbf{x}(n) = z^{-1}\mathbf{e}_j^b(n).$$

3. Verify that the substitutions $\mathbf{z} = \mathbf{y} = \mathbf{x}(n)$, $\mathbf{U} = \mathbf{X}_{1, m}(n)$, and $\mathbf{v} = z^{-m+1}\mathbf{x}(n)$ in (9.4.5b) lead to the forward prediction residual order update in (10.4.3).

4. Give a mathematical justification for the name "partial correlation coefficient" given to the parameter $\Delta_{m+1}(n)$ defined by (10.3.7).

5. Using the association $\mathbf{U} = \mathbf{X}_{1, m}(n)$, $\mathbf{v} = \mathbf{x}(n)$, and $\mathbf{y} = z^{-m-1}\mathbf{x}(n)$ in (9.4.5b) verify that the backward prediction filter error order update is given by (10.3.11).

6. This problem investigates the meaning of the angle parameter $\gamma_m(n)$.
   (a) Verify that the associations $\mathbf{z} = \mathbf{y} = \pi(n)$, $\mathbf{U} = \mathbf{X}_{1, m}(n)$, and $\mathbf{v} = z^{-m-1}\mathbf{x}(n)$ give the angle parameter order update in (10.4.7).
   (b) Suppose that the BPE is non-zero at all N stages of the LS lattice. What does this imply about the "angle" between the data subspaces formed by successively higher order lattice filters?
   (c) Suppose for some m that $\gamma_{m+1}(n) = \gamma_m(n)$. What does this imply about the process that generated the stochastic signal $x(n)$?

REFERENCES

1. M. Morf and D.T. Lee, "Recursive Least Squares Ladder Forms for Fast Parameter Tracking," *Proceedings*, 1978 IEEE Conf. on Dec. and Control, San Diego, CA, January 1978.
2. D.T.L. Lee and M. Morf, "Recursive Square Root Estimation Algorithms," *Proceedings*, 1980 IEEE Int. Conf. on Acous., Speech, and Signal Proc., Denver CO, pp. 1005–1017, April 1980.

3. D.T.L. Lee, M. Morf, and B. Friedlander, "Recursive Least Squares Ladder Estimation Algorithms," *IEEE Transactions on Acous., Speech, and Signal Processing*, vol. ASSP-29, pp. 467–481, June 1981.

4. J.G. Proakis, *Digital Communications*, McGraw-Hill, New York, 1984.

5. B. Friedlander, "Lattice Filters for Adaptive Processing," *Proceedings of the IEEE*, vol. 70, no. 8, pp. 829–867, August, 1982.

6. E.H. Satorius and J. Pack, "Application of Least Squares Lattice Algorithms to Adaptive Equalization," *IEEE Trans. on Communications*, vol. COM-29, pp. 136–142, February 1981.

7. F.Y. Ling and J.G. Proakis, "Generalized Least Squares Lattice Algorithm and its Application to Decision Feedback Equalization," *Proceedings*, 1982 IEEE Int. Conf. on Acous., Speech and Signal Proc., Paris, France, May 1982.

8. F. Ling and J.G. Proakis, "Generalized Multichannel Least Squares Lattice Algorithm Based on Sequential Processing," *IEEE Trans. on Acous., Speech and Signal Processing*, vol. ASSP-32, pp. 381–389, April 1984.

9. W.S. Hodgkiss and J.A. Presley, "Adaptive Tracking of Multiple Sinusoids Whose Power Levels are Widely Separated," *IEEE Trans. on Acous., Speech, and Signal Processing*, vol. ASSP-29, pp. 710–721, June 1981.

10. V.U. Reddy, B. Egardt, and T. Kailath, "Optimized Lattice Form Adaptive Line Enhancer for a Sinusoid Signal in Broadband Noise", *IEEE Trans. on Acous., Speech and Signal Processing*, vol. ASSP-29, pp. 702–710, June 1981.

11. D.T.L. Lee and M. Morf, "A Novel Innovation Based Approach to Pitch Detection," *Proceedings*, 1980 IEEE Int. Conf. on Acous., Speech and Signal Processing, Denver, CO, April 180.

12. J.M. Turner, "Recursive Least Squares Estimation and Lattice Filters," Chapt. 5 in *Adaptive Filters*, C.F.N. Cowan and P.M. Grant, eds., Prentice-Hall, Englewood Cliffs, NJ, 1985.

CHAPTER 11
Fast Transversal Filters

11.1 Introduction

This chapter takes a different approach to fast recursive least squares (RLS) filtering, and applies the projection techniques and vector space methods to the derivation of fixed-order, transversal least squares (LS) filters. The resulting algorithm will be denoted as the fast transversal filter (FTF). The FTF differs from the earlier-derived fast Kalman transversal filter [1, 2] in the following sense. The fast Kalman method derived an efficient algorithm for recursively updating a "correlation" matrix inverse by recognizing the data-shifting property of the correlation matrix. In doing this, a dramatic reduction in operations was achieved compared to the regular RLS method of Chapter 8. For a N-length transversal filter, approximately $10 N$ arithmetic operations per time iteration are required.

However, several of the vector operations of the fast Kalman algorithm may be replaced by simple scalar updates, as was recognized by Carayannis, et al. [3], and this resulted in an LS prediction algorithm requiring approximately $7 N$ operations per time update. A conceptual breakthrough occurred when Cioffi [4], Cioffi and Kailath [5], and Honig [6] realized that the vector space approach described in Chapter 10 for the lattice could also be applied to the efficient computation of exact LS transversal filters and results in the FTF. The FTF method also reduces the required arithmetic operations to approximately $7 N$, but has the added benefit of suggesting geometrical interpretations for the resulting fast transversal algorithm.

This chapter is organized as follows. In Section 11.2, additional vector space concepts needed for the FTF are developed. Section 11.3 then examines the very important transversal filter operator in detail and develops the required LS time update relations for this operator. Section 11.4 then derives the structure of the basic FTF algorithm in detail from the single constraint that the prediction filter be implemented in transversal form. This "basic" FTF has a complexity of $8\ N$ arithmetic operations per time iteration. Lastly, Section 11.5 then illustrates how a simple redefinition of a gain parameter allows the $7\ N$ algorithm to be derived.

It should be mentioned that the FTF does not provide enhanced convergence or tracking performance (compared to least mean squares [LMS]) in all occasions. The paper by Cioffi [9] presents guidelines on when the FTF will be beneficial and when the lower complexity LMS algorithm would be sufficient.

This chapter is meant to be introductory and tutorial concerning the FTF. The FTF and its variants are currently extremely active areas of research in signal processing. Many of its properties, such as stability under finite precision arithmetic implementation and effective initialization techniques, are topics of ongoing investigations. Since these are advanced topics, the reader is referred to the works by Cioffi and Kailath [5, 7, 8] for examples of finite precision and filter initialization considerations.

11.2 Additional Vector Space Relations

Many of the same concepts and results from Chapters 9 and 10 may be used in the current examination of the FTF. One very valuable tool in deriving FTFs is the concept of the transversal filter operator. For the current investigation, this operator is an $N \times n$ matrix, where n is the time index and N is the fixed order of the transversal filter. The resulting FTF algorithm will be seen to be a combination of four separate Nth order transversal filters working in unison. These four filters will be denoted as: (1) the LS prediction filter $\mathbf{w}_N(n)$, (2) the forward prediction error filter $\mathbf{f}_N(n)$, (3) the backward prediction error filter $\mathbf{b}_N(n)$, and (4) the gain filter $\mathbf{g}_N(n)$.

At this point, it is probably not obvious exactly why the operation of the FTF may be partitioned into these four filters. However, in Section 11.4, these filters will be shown to be a direct consequence of: (1) requiring the LS prediction filter $\mathbf{w}_N(n)$ to be transversal in structure, and (2) maintaining the required LS orthogonality conditions at both time $n - 1$ and time n. As will be shown in Section 11.4, these four filters evolve naturally in the course of the LS error minimization. The current section continues with definitions and short investigations of the four transversal filters so that the complete derivation of the time update FTF recursions in Section 11.4 may proceed unimpeded.

Least squares (LS) prediction filter

The general problem of this chapter is again the classical LS problem described in Chapter 9. It is desired to use the LS error criterion to optimally predict the desired signal $d(n)$ using the acquired data signal $x(n)$. An additional requirement now is that the prediction be done with a transversal filter structure. Consider again the previous result (9.3.10) for $\mathbf{w}_N(n)$, the LS transversal filter of length N:

$$\mathbf{w}_N(n) = \langle \mathbf{X}_{0,N-1}(n), \mathbf{X}_{0,N-1}(n) \rangle^{-1} \mathbf{X}_{0,N-1}^T(n)\mathbf{d}(n). \qquad (11.2.1)$$

Recall that this filter $\mathbf{w}_N(n)$ minimizes the norm (squared) of the error between the n-length vector $\mathbf{d}(n)$ and the prediction, $\hat{\mathbf{d}}(n)$, which is constrained to lie in the data subspace $\{\mathbf{X}_{0,N-1}(n)\}$. For the current investigation of transversal LS filters, there is another way to write (11.2.1). For a general $n \times N$ data matrix $\mathbf{U}$, the *transversal filter operator* [4, 5] will be defined as the $N \times n$ matrix

$$\mathbf{K}_U = \langle \mathbf{U}, \mathbf{U} \rangle^{-1}\mathbf{U}^T. \qquad (11.2.2)$$

Thus, by substituting $\mathbf{U} = \mathbf{X}_{0,N-1}(n)$ in (11.2.1), $\mathbf{w}_N(n)$ is immediately given by

$$\mathbf{w}_N(n) = \mathbf{K}_{0,N-1}(n)\mathbf{d}(n). \qquad (11.2.3)$$

The subscripts and time argument on $\mathbf{K}_{0,N-1}(n)$ are consistent with those on the corresponding data matrix. A valuable way of interpreting $\mathbf{K}_U$ is as the "operator" that computes the "best (LS)" filter for predicting a specific signal using a given set of data. Using this interpretation, (11.2.3) may be read as "$\mathbf{K}_{0,N-1}(n)$ computes the filter which gives the LS prediction of $\mathbf{d}(n)$ using the $\mathbf{X}_{0,N-1}(n)$ data matrix." The inner product of the top row of $\mathbf{K}_{0,N-1}(n)$ with $\mathbf{d}(n)$ produces the first coefficient $w_1(n)$, the inner product of the second row of $\mathbf{K}_{0,N-1}(n)$ with $\mathbf{d}(n)$ produces the second coefficient $w_2(n)$, and so on.

The main benefit of writing the LS filter $\mathbf{w}_N(n)$ using the transversal filter operator is that it will provide a straightforward manner for updating $\mathbf{w}_N(n-1)$ to $\mathbf{w}_N(n)$. Note at time $n-1$ that (11.2.3) would give

$$\mathbf{w}_N(n-1) = \mathbf{K}_{0,N-1}(n-1)\mathbf{d}(n-1). \qquad (11.2.4)$$

Since $\mathbf{d}(n)$ is easily obtained from $\mathbf{d}(n-1)$, the FTF algorithm really reduces to finding an efficient method of generating $\mathbf{K}_{0,N-1}(n)$ from a knowledge of $\mathbf{K}_{0,N-1}(n-1)$. This will be accomplished in Section 11.3 using the update relations of Chapter 9.

Several error quantities will be of interest in the later derivations. If $\hat{\mathbf{d}}(n)$ is computed using the transversal filter $\mathbf{w}_N(n)$, then from (9.3.2) and (11.2.3)

$$\mathbf{e}(n|n) = \mathbf{d}(n) - \mathbf{X}_{0,N-1}(n)\mathbf{K}_{0,N-1}(n)\mathbf{d}(n),$$

or using (11.2.1)

$$\mathbf{e}(n|n) = \mathbf{P}_{0,N-1}^{\perp}(n)\mathbf{d}(n). \qquad (11.2.5)$$

Since all error quantities in this chapter are the result of Nth order filters, no order subscript is necessary on the error. Additionally, the current time value of the scalar LS prediction error is easily computed via the inner product

$$e(n|n) = \langle \pi(n), \mathbf{e}(n|n) \rangle = \langle \pi(n), \mathbf{P}_{0,N-1}^{\perp}(n)\mathbf{d}(n) \rangle. \qquad (11.2.6)$$

From (11.2.6), it is immediately seen that the scalar LS prediction error is in the form of an inner product as specified by (9.4.5b). This will be very useful, since the results from Chapter 9 for updating an inner product in a Hillbert space will thus apply to transversal filter updates.

Forward prediction error (FPE) filter

The second LS transversal filter used in the FTF algorithm is an Nth order forward linear prediction filter. This filter computes the forward prediction error (FPE) between the current data vector $x(n)$ and a prediction $\hat{x}_f(n)$ based upon a knowledge of past data vectors. To see the structure of this approach, expand (10.2.1) for the individual FPE equations, assuming a transversal implementation:

$$e^f(1|n) = x(1)$$
$$e^f(2|n) = x(2) - f_1(n)x(1)$$

(11.2.7)

$$\cdots \qquad \cdots$$

$$e^f(n|n) = x(n) - f_1(n)x(n-1) - f_2(n)x(n-2) - \cdots - f_N(n)x(n-N),$$

where the $f_i(n)$ are the components of the forward linear prediction filter $\mathbf{f}_N(n)$:

$$\mathbf{f}_N(n) = [f_1(n), f_2(n), \ldots, f_N(n)]^T.$$

(11.2.8)

In vector notation, (11.2.7) becomes

$$\mathbf{e}^f(n|n) = \mathbf{x}(n) - \hat{\mathbf{x}}_f(n),$$

(11.2.9)

where the forward prediction $\hat{\mathbf{x}}_f(n)$ of the current data vector $\mathbf{x}(n)$ is given by

$$\hat{\mathbf{x}}_f(n) = \mathbf{X}_{1,N}(n)\mathbf{f}_N(n),$$

(11.2.10)

and $\mathbf{X}_{1,N}(n)$ follows the delay notation of Chapter 9:

$$\mathbf{X}_{1,N}(n) = [z^{-1}\mathbf{x}(n), \ldots, z^{-N}\mathbf{x}(n)].$$

(11.2.11)

However, using the concept of the projection matrix, the LS prediction of $\mathbf{x}(n)$, which minimizes $\langle \mathbf{e}^f(n|n), \mathbf{e}^f(n|n) \rangle$, is given simply as

$$\hat{\mathbf{x}}_f(n) = \mathbf{P}_{1,N}(n)\mathbf{x}(n),$$

(11.2.12)

and the FPE vector becomes

$$\mathbf{e}^f(n|n) = \mathbf{P}_{1,N}^\perp(n)\mathbf{x}(n).$$

(11.2.13)

Therefore, using the definition of a projection matrix from (9.3.13), and incorporating (11.2.2) for the transversal filter operator gives

$$\mathbf{f}_N(n) = \mathbf{K}_{1,N}(n)\mathbf{x}(n).$$

(11.2.14)

Since (11.2.14) holds for any time value, it is also true that

$$\mathbf{f}_N(n-1) = \mathbf{K}_{1,N}(n-1)\mathbf{x}(n-1).$$

(11.2.15)

Therefore, $\mathbf{f}_N(n)$ could be recursively updated if $\mathbf{K}_{1,N}(n-1)$ could be recursively updated to produce $\mathbf{K}_{1,N}(n)$. This procedure will be examined in Section 11.3.

Similar to (11.2.6), the current FPE sample $e^f(n|n)$ is given by the inner product

$$e^f(n|n) = \langle \pi(n), e^f(n|n) \rangle = \langle \pi(n), \mathbf{P}_{1,N}^{\perp}(n)\mathbf{x}(n) \rangle, \qquad (11.2.16)$$

where (11.2.14) has also been used. Thus, the scalar FPE in (11.2.16) may be written as an inner product of the form specified by (9.4.5b). Additionally, the FPE residual or the energy of the FPE vector $e^f(n|n)$ is given by

$$\varepsilon^f(n) = \langle e^f(n|n), e^f(n|n) \rangle = \langle \mathbf{x}(n), \mathbf{P}_{1,N}^{\perp}(n)\mathbf{x}(n) \rangle, \qquad (11.2.17)$$

which is easily derived using the definition (11.2.13). Consistent with much of the current literature, this is called the Nth order FPE residual at time n. From (11.2.17), note that $\varepsilon^f(n)$ is now also in the form required by (9.4.5b) for updating of an inner product in a Hilbert space.

Backward prediction filter

The third transversal filter necessary in the FTF algorithm is an Nth order backward prediction filter. Following a development similar to that for the FPE, the backward prediction error (BPE) equations of (10.2.7) may be expanded to obtain

$$e^b(1|n) = 0 - b_1(n)x(1)$$

$$\cdots$$

$$e^b(N|n) = 0 - b_1(n)x(N) - \cdots - b_N(n)x(1)$$
$$\qquad (11.2.18)$$

$$\cdots$$

$$e^b(n-1|n) = x(n-N-1) - b_1(n)x(n-1) - \cdots - b_N(n)x(n-N)$$

$$e^b(n|n) = x(n-N) - b_1(n)x(n) - \cdots - b_N(n)x(n-N+1).$$

Let the backward prediction filter be defined as the vector of the $b_i(n)$:

$$\mathbf{b}_N(n) = [b_1(n), b_2(n), \ldots, b_N(n)]^T. \qquad (11.2.19)$$

In vector form, the backward error equations of (11.2.18) may now be written as

$$e^b(n|n) = z^{-N}\mathbf{x}(n) - \hat{\mathbf{x}}_b(n-N), \qquad (11.2.20)$$

where the backward prediction, $\hat{\mathbf{x}}_b(n-N)$, of the delayed data vector $z^{-N}\mathbf{x}(n)$ is given by

$$\hat{\mathbf{x}}_b(n-N) = \mathbf{X}_{0,N-1}(n)\mathbf{b}_N(n). \qquad (11.2.21)$$

In (11.2.21), $\mathbf{X}_{0,N-1}(n)$ is the data matrix used by the backward predictor, defined using the notation of Chapter 9:

$$\mathbf{X}_{0,N-1}(n) = [\mathbf{x}(n), z^{-1}\mathbf{x}(n), \ldots, z^{-N+1}\mathbf{x}(n)]. \qquad (11.2.22)$$

With the data matrix now defined by (11.2.22), the solution for the LS backward prediction is immediately given by

$$\hat{\mathbf{x}}_b(n - N) = \mathbf{P}_{0,N-1}(n)z^{-N}\mathbf{x}(n), \qquad (11.2.23)$$

and the LS BPE vector,

$$\mathbf{e}^b(n|n) = \mathbf{P}_{0,N-1}^{\perp}(n)z^{-N}\mathbf{x}(n). \qquad (11.2.24)$$

Equating (11.2.21) and (11.2.23) therefore gives

$$\mathbf{b}_N(n) = \langle \mathbf{X}_{0,N-1}(n), \mathbf{X}_{0,N-1}(n) \rangle^{-1}\mathbf{X}_{0,N-1}^T(n)z^{-N}\mathbf{x}(n). \qquad (11.2.25)$$

However, using the definition of the transversal filter operator from (11.2.2) with $\mathbf{U} = \mathbf{X}_{0,N-1}(n)$ gives the desired form

$$\mathbf{b}_N(n) = \mathbf{K}_{0,N-1}(n)z^{-N}\mathbf{x}(n). \qquad (11.2.26)$$

In a manner similar to that for the scalar FPE, the scalar BPE, $e^b(n|n)$, is given simply by

$$e^b(n|n) = \langle \pi(n), \mathbf{e}^b(n|n) \rangle = \langle \pi(n), \mathbf{P}_{0,N-1}^{\perp}(n)z^{-N}\mathbf{x}(n) \rangle. \qquad (11.2.27)$$

Thus, the BPE can be written as an inner product of the form (9.4.5b). Similar to the FPE residual, the computation of the BPE residual is needed in the derivation of the FTF. The Nth order BPE residual is given by

$$\varepsilon^b(n) = \langle \mathbf{e}^b(n|n), \mathbf{e}^b(n|n) \rangle = \langle z^{-N}\mathbf{x}(n), \mathbf{P}_{0,N-1}^{\perp}(n)z^{-N}\mathbf{x}(n) \rangle, \qquad (11.2.28)$$

which is easily derived using the definition (11.2.24).

Gain transversal filter

The last of the four transversal filters necessary to implement the FTF algorithm will be referred to as the *gain transversal filter* $\mathbf{g}_N(n)$. When examined from the standpoint of a transversal implementation, $\mathbf{g}_N(n)$ has an especially important interpretation. Consider Figure 11.1 displaying the geometry of the one-dimensional subspace $\mathbf{x}(n)$ as it changes from time $n - 1$ to time n. From (9.5.12), the angle parameter quantifying this change is

$$\gamma_1(n) = \cos^2 \theta.$$

The transversal gain filter will be seen to provide another way of quantifying this angular change within the framework of a transversal LS filter estimate.

To see this, consider the projection of $\pi(n)$ onto $\mathbf{x}(n)$, which would be denoted by $\mathbf{P}_\mathbf{x}(n)\pi(n)$ in the vector space formulation. Since this projection in Figure 11.1 is in the direction of $\mathbf{x}(n)$, it can also be written as $g(n)\mathbf{x}(n)$, where $g(n)$ is a scalar multiplier valid at time n. Therefore,

$$g(n)\mathbf{x}(n) = \mathbf{P}_\mathbf{x}(n)\pi(n). \qquad (11.2.29)$$

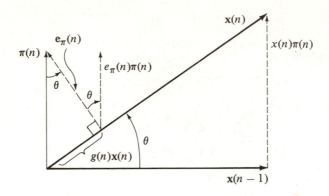

Figure 11.1   Construction of optimal LS gain filter $g(n)$ for the one coefficient case.

Thus, the vector $g(n)\mathbf{x}(n)$ is the LS prediction of the unit time vector $\pi(n)$ using the data $\mathbf{x}(n)$. Equivalently, $g(n)$ is the optimal one coefficient LS filter for predicting $\pi(n)$ using the data $\mathbf{x}(n)$. The error vector $\mathbf{e}_{\pi}(n)$ resulting from this prediction is

$$\mathbf{e}_{\pi}(n) = \pi(n) - \mathbf{P}_{\mathbf{x}}(n)\pi(n) = \mathbf{P}_{\mathbf{x}}^{\perp}(n)\pi(n),$$

and the scalar current time n component of the error vector is therefore

$$e_{\pi}(n) = \langle \pi(n), \mathbf{P}_{\mathbf{x}}^{\perp}(n)\pi(n) \rangle = \gamma_1(n), \tag{11.2.30}$$

where the definition (9.5.12) has been used. Combining these results then produces

$$\gamma_1(n) = \cos^2 \theta = 1 - \langle \pi(n), \mathbf{x}(n)g(n) \rangle, \tag{11.2.31}$$

from which

$$\sin^2 \theta = \langle \pi(n), \mathbf{x}(n)g(n) \rangle = \mathbf{x}(n)g(n).$$

Thus, the gain filter $g(n)$ is also seen to quantify the subspace angular change. This is important since $g(n)$ can be interpreted as another LS transversal filter.

To see this, consider the extension to a general N dimension subspace $\mathbf{X}_{0,N-1}(n)$. Analogous to (11.2.29),

$$\mathbf{X}_{0,N-1}(n)\mathbf{g}_N(n) = \mathbf{P}_{0,N-1}(n)\pi(n) \tag{11.2.32}$$

analogous to (11.2.30),

$$\gamma_N(n) = \langle \pi(n), \mathbf{P}_{0,N-1}^{\perp}(n)\pi(n) \rangle \tag{11.2.33}$$

and analogous to (11.3.31)

$$\gamma_N(n) = 1 - \mathbf{x}_N^T(n)\mathbf{g}_N(n), \tag{11.2.34}$$

where

$$\mathbf{x}_N(n) = [x(n), x(n-1), \ldots, x(n-N+1)]^T. \tag{11.2.35}$$

The main result of this is that from (11.2.34), the angular parameter $\gamma_N(n)$ can be computed using the output of the gain transversal filter $\mathbf{g}_N(n)$ operating on the N most recent samples of the data. Furthermore, using (11.2.32), $\mathbf{g}_N(n)$ can be written immediately as

$$\mathbf{g}_N(n) = \mathbf{K}_{0,N-1}(n)\pi(n), \tag{11.2.36}$$

which defines the gain filter as the "LS predictor of $\pi(n)$ using the data matrix $\mathbf{X}_{0,N-1}(n)$."

Now that the four transversal filters have been defined, the derivation of the FTF algorithm may proceed. Since the transversal filter operator has been seen to be pivotal in computing LS filters, a method for time updating the transversal operator must be determined. This is the topic of the following section.

11.3 The Transversal Filter Operator Update

The derivation of the FTF algorithm consists of determining: (1) the time updates for the four transversal filters of Section 11.2, and (2) the manner in which the four filters interact to produce the desired LS algorithm. This section examines topic (1), and topic (2) is the subject of Section 11.4. The pivotal result of this section will be deriving the time updates for the transversal filter operators $\mathbf{K}_{0,N-1}(n)$ and $\mathbf{K}_{1,N}(n)$, which were shown in Section 11.2 to be necessary for defining the four LS transversal filters. The starting point is (9.4.3a) for the projection matrix update, repeated here for a general vector space $\{\mathbf{U}\}$ and a general n-component vector $\mathbf{v}$:

$$\mathbf{P}_{\mathbf{Uv}} = \mathbf{P}_{\mathbf{U}} + \mathbf{P}_{\mathbf{U}}^{\perp}\mathbf{v}\langle\mathbf{P}_{\mathbf{U}}^{\perp}\mathbf{v}, \mathbf{P}_{\mathbf{U}}^{\perp}\mathbf{v}\rangle^{-1}\mathbf{v}^T\mathbf{P}_{\mathbf{U}}^{\perp}. \tag{11.3.1}$$

In (11.3.1), it has been assumed that $\mathbf{v}$ is appended as the last column of the matrix $[\mathbf{U}, \mathbf{v}]$. It will be seen later that the ordering of $\mathbf{U}$ and $\mathbf{v}$ does make a difference in the resulting transversal filter operator update. For now, consider the effects of premultiplying (11.3.1) by the transversal filter operator for the $\{\mathbf{U}, \mathbf{v}\}$ space, $\mathbf{K}_{\mathbf{Uv}}$:

$$\mathbf{K}_{\mathbf{Uv}}\mathbf{P}_{\mathbf{Uv}} = \mathbf{K}_{\mathbf{Uv}}\mathbf{P}_{\mathbf{U}} + \mathbf{K}_{\mathbf{Uv}}\mathbf{P}_{\mathbf{U}}^{\perp}\mathbf{v}\langle\mathbf{P}_{\mathbf{U}}^{\perp}\mathbf{v}, \mathbf{P}_{\mathbf{U}}^{\perp}\mathbf{v}\rangle^{-1}\mathbf{v}^T\mathbf{P}_{\mathbf{U}}^{\perp}. \tag{11.3.2}$$

Since every row of $\mathbf{K}_{\mathbf{Uv}}$ is a vector in the vector space $\{\mathbf{U}, \mathbf{v}\}$, then post-multiplying each row of $\mathbf{K}_{\mathbf{Uv}}$ by the projection matrix $\mathbf{P}_{\mathbf{Uv}}$ simply reproduces the original rows of $\mathbf{K}_{\mathbf{Uv}}$. Thus,

$$\mathbf{K}_{\mathbf{Uv}}\mathbf{P}_{\mathbf{Uv}} = \mathbf{K}_{\mathbf{Uv}}. \tag{11.3.3}$$

To simplify the first term on the right side of (11.3.2), use the definition of $\mathbf{K}_{\mathbf{Uv}}$ from (11.2.2):

$$\mathbf{K}_{\mathbf{Uv}} = \langle[\mathbf{U}, \mathbf{v}], [\mathbf{U}, \mathbf{v}]\rangle^{-1}[\mathbf{U}, \mathbf{v}]^T. \tag{11.3.4}$$

Hence, postmultiplication of (11.3.4) by $[\mathbf{U}, \mathbf{v}]$ gives

$$\mathbf{K}_{\mathbf{U}\mathbf{v}}[\mathbf{U}, \mathbf{v}] = \mathbf{I}. \tag{11.3.5}$$

Since $\mathbf{K}_{\mathbf{U}\mathbf{v}}$ is an $(N + 1) \times (n)$ matrix, $\mathbf{U}$ is an $n \times N$ matrix, and $\mathbf{v}$ is an $n \times 1$ vector, (11.3.5) may thus be written as the set of partitioned equations

$$\mathbf{K}_{\mathbf{U}\mathbf{v}}\mathbf{U} = \begin{bmatrix} \mathbf{I}_{NN} \\ 0_N^T \end{bmatrix} \tag{11.3.6}$$

and

$$\mathbf{K}_{\mathbf{U}\mathbf{v}}\mathbf{v} = \begin{bmatrix} 0_N \\ 1 \end{bmatrix}, \tag{11.3.7}$$

where $\mathbf{I}_{NN}$ is the $N \times N$ identity matrix and 0_N is the N-length null vector. Therefore, since

$$\mathbf{K}_{\mathbf{U}\mathbf{v}}\mathbf{P}_{\mathbf{U}} = \mathbf{K}_{\mathbf{U}\mathbf{v}}\mathbf{U}\langle\mathbf{U}, \mathbf{U}\rangle^{-1}\mathbf{U}^T,$$

then using (11.3.6) immediately produces

$$\mathbf{K}_{\mathbf{U}\mathbf{v}}\mathbf{P}_{\mathbf{U}} = \begin{bmatrix} \mathbf{I}_{NN} \\ 0_N^T \end{bmatrix}\mathbf{K}_{\mathbf{U}} = \begin{bmatrix} \mathbf{K}_{\mathbf{U}} \\ 0_n^T \end{bmatrix}. \tag{11.3.8}$$

Using (11.3.3)–(11.3.8) in (11.3.2) then gives the following form:

$$\mathbf{K}_{\mathbf{U}\mathbf{v}} = \begin{bmatrix} \mathbf{K}_{\mathbf{U}} \\ 0_n^T \end{bmatrix} + \left\{ \begin{bmatrix} 0_N \\ 1 \end{bmatrix} - \begin{bmatrix} \mathbf{K}_{\mathbf{U}}\mathbf{v} \\ 0 \end{bmatrix} \right\}\langle\mathbf{P}_{\mathbf{U}}^{\perp}\mathbf{v}, \mathbf{P}_{\mathbf{U}}^{\perp}\mathbf{v}\rangle^{-1}\mathbf{v}^T\mathbf{P}_{\mathbf{U}}^{\perp}. \tag{11.3.9a}$$

Note that the ordering of $\mathbf{U}$ and $\mathbf{v}$ does make a subtle but important difference. Suppose $\mathbf{v}$ were appended to the first column to form the matrix $[\mathbf{v}, \mathbf{U}]$. Then it is straightforward to show that

$$\mathbf{K}_{\mathbf{v}\mathbf{U}} = \begin{bmatrix} 0_n^T \\ \mathbf{K}_{\mathbf{U}} \end{bmatrix} + \left\{ \begin{bmatrix} 1 \\ 0_N \end{bmatrix} - \begin{bmatrix} 0 \\ \mathbf{K}_{\mathbf{U}}\mathbf{v} \end{bmatrix} \right\}\langle\mathbf{P}_{\mathbf{U}}^{\perp}\mathbf{v}, \mathbf{P}_{\mathbf{U}}^{\perp}\mathbf{v}\rangle^{-1}\mathbf{v}^T\mathbf{P}_{\mathbf{U}}^{\perp} \tag{11.3.9b}$$

A judicious choice of the $\mathbf{U}$ matrix and $\mathbf{v}$ vector in (11.3.9) allows time update recursions to be derived for the transversal filter operators $\mathbf{K}_{1,N}(n)$ and $\mathbf{K}_{0,N-1}(n)$. In each case, the $\mathbf{v}$ vector is chosen to be the unit vector $\pi(n)$. For the case of the $\mathbf{K}_{0,N-1}(n)$ operator, the $\{\mathbf{U}\}$ subspace is $\{\mathbf{X}_{0,N-1}(n)\}$, whereas the $\mathbf{K}_{1,N}(n)$ update utilizes the $\{\mathbf{X}_{1,N}(n)\}$ subspace.

As an example, consider the case for $\mathbf{K}_{0,N-1}(n)$. Using the z-shift notation, the vector space $\{\mathbf{U}, \mathbf{v}\}$ is defined by its basis vectors as

$$\{\mathbf{U}, \mathbf{v}\} = \{\mathbf{x}(n), z^{-1}\mathbf{x}(n), \ldots, z^{-N+1}\mathbf{x}(n), \pi(n)\},$$

which may be written compactly as

$$\{\mathbf{U}, \mathbf{v}\} = \{\mathbf{X}_{0,N-1}(n), \pi(n)\}. \tag{11.3.10}$$

Therefore,

$$\mathbf{K}_{\mathbf{U}\mathbf{v}} = \mathbf{K}_{\mathbf{U}, \pi} = \mathbf{K}_{0,N-1,\pi}(n), \tag{11.3.11}$$

$$\mathbf{P}_{\mathbf{U}\mathbf{v}} = \mathbf{P}_{\mathbf{U},\pi} = \mathbf{P}_{0,N-1,\pi}(n), \tag{11.3.12}$$

where the notation "0, $N - 1$, π" denotes using the vector space created by appending $\pi(n)$ after the final column of $\mathbf{X}_{0,N-1}(n)$, as in (11.3.10). However, (11.3.12) is simply the time decomposition for the projection matrix derived from (9.5.5). Therefore,

$$\mathbf{P}_{0,N-1,\pi}(n) = \begin{bmatrix} \mathbf{P}_{0,N-1}(n-1) & \mathbf{0}_{n-1} \\ \mathbf{0}_{n-1}^T & 1 \end{bmatrix}. \tag{11.3.13}$$

Using (11.3.13) in (11.3.3), it is easy to show that the following form for $\mathbf{K}_{0,N-1,\pi}(n)$ satisfies the relation (11.2.2):

$$\mathbf{K}_{0,N-1,\pi}(n) = \begin{bmatrix} \mathbf{K}_{0,N-1}(n-1) & \mathbf{0}_N^T \\ \mathbf{c}^T(n-1) & 1 \end{bmatrix}, \tag{11.3.14}$$

where $\mathbf{c}(n-1)$ is an $n-1$ component vector not used in the ensuing FTF derivations.

In a development similar to the preceding, appending the time selection vector $\pi(n)$ to the $\mathbf{X}_{1,N}(n)$ matrix allows time decompositions of the transversal filter operator $\mathbf{K}_{1,N,\pi}(n)$. The decomposition for $\mathbf{K}_{1,N,\pi}(n)$ is derived in a manner similar to (11.3.14) and is given by

$$\mathbf{K}_{1,N,\pi}(n) = \begin{bmatrix} \mathbf{K}_{1,N}(n-1) & \mathbf{0}_N \\ \mathbf{b}^T(n-1) & 1 \end{bmatrix}. \tag{11.3.15}$$

where $\mathbf{b}(n-1)$ is another $n-1$ component vector that is not used explicitly in the ensuing derivations.

At this point, all preliminary definitions and relations needed to derive the time updates for the four transversal filters have been developed. Section 11.4 next provides a detailed derivation of the FTF algorithm.

11.4 The FTF Time Updates

To help understand the FTF derivation, the following approach is used in this text: (1) first, require the LS prediction filter for predicting $d(n)$ using $x(n), \ldots, x(n-N+1)$ to be implemented in a transversal filter structure, and (2) examine the conditions that result from enforcing the necessary LS orthogonality conditions. The resulting equations will be seen to require another set of updates that turn out to be the gain filter updates. Fulfilling this set of updates will, in turn, require the updates of the FPE and BPE filters. Utilizing this method of working "back to the start," the need for each of the four filters will become explicit displayed. Additionally, it will be seen that these four filters, as well as other scalar parameters, are all a natural consequence of minimizing the original LS error criterion.

LS filter update

The overall objective of the FTF algorithm is to efficiently update the original LS filter $\mathbf{w}_N(n)$. In terms of the transversal filter operator from (11.2.3),

$$\mathbf{w}_N(n) = \mathbf{K}_{0,N-1}(n)\mathbf{d}(n). \tag{11.4.1}$$

It is desired to compute $\mathbf{w}_N(n)$ by updating the previous value $\mathbf{w}_N(n-1)$, which thus requires updating the transversal filter operator.

To derive the corresponding LS prediction filter update, make the substitutions $\mathbf{U} = \mathbf{X}_{0,N-1}(n)$ and $\mathbf{v} = \pi(n)$ in (11.3.9a), and use the partitioning of (11.3.14) for $\mathbf{K}_{0,N-1,\pi}(n)$. Postmultiplying the result by $\mathbf{d}(n)$ gives

$$
\begin{bmatrix} \mathbf{K}_{0,N-1}(n-1) & \mathbf{0}_N \\ \mathbf{c}^T(n-1) & 1 \end{bmatrix} \begin{bmatrix} \mathbf{d}(n-1) \\ d(n) \end{bmatrix}
$$
$$
= \begin{bmatrix} \mathbf{K}_{0,N-1}(n) \\ 0 \end{bmatrix} \mathbf{d}(n) - \begin{bmatrix} \mathbf{g}_N(n) \\ -1 \end{bmatrix} \frac{\langle \pi(n), \mathbf{P}_{0,N-1}^{\perp}(n)\mathbf{d}(n) \rangle}{\gamma_N(n)}, \tag{11.4.2}
$$

where (11.2.33) and (11.2.36) have also been used. Expanding the top partition of (11.4.2) and using (11.2.3) and (11.2.5) then gives

$$\mathbf{w}_N(n) = \mathbf{w}_N(n-1) + \frac{e(n|n)}{\gamma_N(n)}\mathbf{g}_N(n). \tag{11.4.3}$$

Equation (11.4.3) states a very important result; namely, that the optimal LS filter at time n can be computed from the "old" LS filter at time $n-1$ plus an update vector that is proportional to the gain filter. The old filter $\mathbf{w}_N(n-1)$ was computed at the previous iteration. However, the $e(n|n)$, $\gamma_N(n)$, and $\mathbf{g}_N(n)$ must all be computed during the present iteration before $\mathbf{w}_N(n)$ can be calculated. Therefore, derivation of the FTF algorithm consists of determining the LS updates for the unknown LS parameters $e(n|n)$, $\gamma_N(n)$, and $\mathbf{g}_N(n)$.

The strategy of the FTF derivation is thus to use the appropriate LS recursions to update these unknown LS parameters at time n based upon their known values at time $n-1$. In so doing, however, other LS parameters may be introduced. If so, these new LS parameters must also be computable in a recursive fashion for the resulting algorithm to be implementable.

Gain filter update

To compute an update for $\mathbf{g}_N(n)$ in terms of $\mathbf{g}_N(n-1)$, make the substitutions $\mathbf{U} = \mathbf{X}_{1,N}(n)$, $\mathbf{v} = \mathbf{x}(n)$ in (11.3.9b), such that $\{\mathbf{v}, \mathbf{U}\} = \{\mathbf{X}_{0,N}(n)\}$, and postmultiply the result by $\pi(n)$. Doing this and using (11.2.16) and (11.2.17) in the result gives

$$\mathbf{K}_{0,N}(n)\pi(n) = \begin{bmatrix} \mathbf{0}_n^T \\ \mathbf{K}_{1,N}(n) \end{bmatrix}\pi(n) + \begin{bmatrix} 1 \\ -\mathbf{K}_{1,N}(n)\mathbf{x}(n) \end{bmatrix}\frac{e^f(n|n)}{\varepsilon^f(n)}. \tag{11.4.4}$$

The left-hand side of (11.4.4) reads "the N + 1st order LS predictor of $\pi(n)$ using $\mathbf{X}_{0,N}(n)$," which, consistent with (11.2.36), must be the N + 1st order gain filter $\mathbf{g}_{N+1}(n)$. It will be very useful in the following work to partition $\mathbf{g}_{N+1}(n)$ into

$$\mathbf{g}_{N+1}(n) = \begin{bmatrix} \mathbf{k}_N(n) \\ k(n) \end{bmatrix}, \tag{11.4.5}$$

where $\mathbf{k}_N(n)$ is the vector of the first N components of $\mathbf{g}_{N+1}(n)$ and the scalar $k(n)$ is the N + 1st coefficient. It is also easy to show by direct substitution in (11.2.36) that

$$\mathbf{g}_N(n - 1) = \mathbf{K}_{1,N}(n)\pi(n). \tag{11.4.6}$$

Using these results in (11.4.4) gives

$$\mathbf{g}_{N+1}(n) = \begin{bmatrix} \mathbf{k}_N(n) \\ k(n) \end{bmatrix} = \begin{bmatrix} 0 \\ \mathbf{g}_N(n-1) \end{bmatrix} + \frac{e^f(n|n)}{\varepsilon^f(n)} \begin{bmatrix} 1 \\ -\mathbf{f}_N(n) \end{bmatrix}. \tag{11.4.7}$$

Note in (11.4.7) that additional new LS variables $(e^f(n|n), \varepsilon^f(n), \mathbf{f}_N(n))$ related to the forward prediction problem have been introduced. Thus, the concept of a forward prediction filter "embedded" within the overall LS prediction problem is seen to naturally evolve as a consequence of the gain filter update. Moreover, since all the new parameters are a function of current time n, they must all be computed before the update in (11.4.7) may be done. Also, note in (11.4.7) that the optimal N coefficient gain filter at time $n - 1$ is explicitly displayed on the right-hand side. This gain has been computed at iteration $n - 1$, and therefore is currently available. On the left-hand side, however, the preceding has produced the optimal $N + 1$ coefficient gain filter at time n, instead of the optimal N coefficient gain filter. It must be emphasized that, in general, the first N coefficients of an optimal $N + 1$ length transversal filter will not necessarily be equivalent to those of the optimal N-length transversal filter.

One way to benefit from (11.4.7) is to compute an alternative expression for $\mathbf{g}_{N+1}(n)$ as a function of $\mathbf{g}_N(n)$, and then equate it to $\mathbf{g}_{N+1}(n)$ from (11.4.7). To do this, make the substitutions $\mathbf{U} = \mathbf{X}_{0,N-1}(n)$, $\mathbf{v} = z^{-N}\mathbf{x}(n)$ in (11.3.9a), such that $\{\mathbf{U}, \mathbf{v}\} = \{\mathbf{X}_{0,N}(n)\}$, and postmultiply the result by $\pi(n)$, obtaining

$$\begin{bmatrix} \mathbf{g}_N(n) \\ 0 \end{bmatrix} = \begin{bmatrix} \mathbf{k}_N(n) \\ k(n) \end{bmatrix} + \begin{bmatrix} \mathbf{b}_N(n) \\ -1 \end{bmatrix} \frac{e^b(n|n)}{\varepsilon^b(n)}, \tag{11.4.8}$$

where (11.2.26) and (11.2.27) have also been used for the backward error parameters. One result that is important for work to come is that the lower partition of (11.4.8) provides

$$k(n) = \frac{e^b(n|n)}{\varepsilon^b(n)}. \tag{11.4.9}$$

But this variable $k(n)$ has been previously computed as the final component of $\mathbf{g}_{N+1}(n)$ in the update (11.4.7). Therefore, the top partition of (11.4.8) may be simplified to

$$\mathbf{g}_N(n) = \mathbf{k}_N(n) + k(n)\mathbf{b}_N(n). \tag{11.4.10}$$

Assuming the update (11.4.7) has been completed, then (11.4.10) shows that the desired update $\mathbf{g}_N(n)$ may be computed in terms of $\mathbf{g}_{N+1}(n)$. However, similar to the set of FPE variables, a set of BPE parameters $(e^b(n|n), \varepsilon^b(n), \mathbf{b}_N(n))$ have been introduced in (11.4.8). Since they are all functions of current time, they must be computed before (11.4.8) can be used. It should be kept in mind that these FPE and BPE variables appear solely as a result of computing the optimal LS gain $\mathbf{g}_N(n)$, as required by the LS filter update (11.4.3) for $\mathbf{w}_N(n)$. Therefore, before continuing with the gain update derivation, it will be necessary to develop the LS update relations for the FPE and BPE variables.

FPE filter update

Note from (11.4.7) that $\mathbf{g}_{N+1}(n)$ is given in terms of $\mathbf{f}_N(n)$, the FPE filter at time n. Since $\mathbf{f}_N(n)$ at time n is not yet available at this step, the previous filter $\mathbf{f}_N(n-1)$ must be updated to the required value. Using (11.2.14) and the top partition in (11.3.9a), let $\mathbf{U} = \mathbf{X}_{1,N}(n)$, $\mathbf{v} = \pi(n)$, and postmultiply the result by $\mathbf{x}(n)$. Simplifying provides

$$\mathbf{f}_N(n) = \mathbf{f}_N(n-1) + \frac{\mathbf{g}_N(n-1)e^f(n|n)}{\langle \pi(n), \mathbf{P}_{1,N}^{\perp}(n)\pi(n)\rangle}, \tag{11.4.11}$$

where (11.4.6) has also been used. However, by direct substitution, it is a simple matter to show from (11.2.33) that

$$\gamma_N(n-1) = \langle \pi(n), \mathbf{P}_{1,N}^{\perp}(n)\pi(n)\rangle. \tag{11.4.12a}$$

Using (11.4.12a) in (11.4.11) then provides

$$\mathbf{f}_N(n) = \mathbf{f}_N(n-1) + \frac{e^f(n|n)}{\gamma_N(n-1)}\mathbf{g}_N(n-1). \tag{11.4.12b}$$

Since $\mathbf{g}_N(n-1)$ and $\gamma_N(n-1)$ have time arguments "$n-1$," they are available from the previous iteration. However, $e^f(n|n)$ is the "FPE from predicting $x(n)$ using the updated filter $\mathbf{f}_N(n)$." Therefore, $e^f(n|n)$ must be computed before the updates for $\mathbf{f}_N(n)$ may be done.

FPE update

To compute $e^f(n|n)$, first consider its definition. It is the nth component of the vector $e^f(n|n)$, and from (11.2.7) is therefore given by

$$e^f(n|n) = x(n) - \mathbf{x}_N^T(n-1)\mathbf{f}_N(n), \tag{11.4.13}$$

where (11.2.35) defines $\mathbf{x}_N(n-1)$. Substituting (11.4.12b) into (11.4.13) then provides

$$e^f(n|n) = e^f(n|n-1) - \frac{e^f(n|n)}{\gamma_N(n-1)} \mathbf{x}_N^T(n-1)\mathbf{g}_N(n-1), \quad (11.4.14)$$

where

$$e^f(n|n-1) = x(n) - \mathbf{x}_N^T(n-1)\mathbf{f}_N(n-1) \quad (11.4.15)$$

is the "FPE from predicting $x(n)$ using the filter at time $n-1$." Solving (11.4.14) for $e^f(n|n)$ and using (11.2.34) then produces

$$e^f(n|n) = \gamma_N(n-1)e^f(n|n-1). \quad (11.4.16)$$

Both of the parameters on the right-hand side of (11.4.16) are currently available, and therefore $e^f(n|n)$ may be computed.

FPE residual update

The remaining unknown parameter in (11.4.7) is the FPE residual. Its update may be accomplished by using (9.4.5b) for the update of the inner product $\langle \mathbf{z}, \mathbf{P}_{\mathbf{U}\mathbf{v}}^{\perp}\mathbf{y} \rangle$ with $\mathbf{U} = \mathbf{X}_{1,N}(n)$, $\mathbf{v} = \pi(n)$, and $\mathbf{z} = \mathbf{y} = \mathbf{x}(n)$. Using (9.5.5) to time partition the resulting orthogonal projection matrix $\mathbf{P}_{1,N,\pi}^{\perp}(n)$ and using (11.2.16) and (11.2.17) then gives the following update:

$$\varepsilon^f(n) = \varepsilon^f(n-1) + \frac{[e^f(n|n)]^2}{\gamma_N(n-1)}. \quad (11.4.17)$$

Substituting (11.4.16) for one of the $e^f(n|n)$ terms in (11.4.17) then provides

$$\varepsilon^f(n) = \varepsilon^f(n-1) + e^f(n|n)e^f(n|n-1). \quad (11.4.18)$$

BPE filter update

The BPE filter derivations are similar to those of the FPE and are only outlined here. First, use the top partition in (11.3.9a) with $\mathbf{U} = \mathbf{X}_{0,N-1}(n)$, and $\mathbf{v} = \pi(n)$, and postmultiply by $z^{-N}x(n)$. Then simplifying the result using (11.2.27), (11.2.33), and (11.2.36) gives the desired form for the update

$$\mathbf{b}_N(n) = \mathbf{b}_N(n-1) + \mathbf{g}_N(n)\frac{e^b(n|n)}{\gamma_N(n)}. \quad (11.4.19)$$

Therefore, to update $\mathbf{b}_N(n)$, it is seen that the updated angle parameter $\gamma_N(n)$ must also be calculated. This update will be derived after first completing the remaining backward error updates.

BPE update

To update $e^b(n|n)$ for use in (11.4.9), simply premultiply (11.4.19) by $\mathbf{x}_N^T(n)$ and subtract the result from the sample $x(n-N)$, giving

$$e^b(n|n) = e^b(n|n-1) + \langle \mathbf{x}_N(n), \mathbf{g}_N(n) \rangle \frac{e^b(n|n)}{\gamma_N(n)}. \quad (11.4.20)$$

Using (11.2.34) in the above and simplifying then gives

$$e^b(n|n) = \gamma_N(n)e^b(n|n-1). \tag{11.4.21}$$

One method of computing $e^b(n|n-1)$ is to use the definition

$$e^b(n|n-1) = x(n-N) - \mathbf{x}_N^T(n)\mathbf{b}_N(n-1). \tag{11.4.22}$$

This computation will be used in the basic FTF algorithm of this section. However, it will be seen in Section 11.5 that there is a scalar operation for computing $e^b(n|n-1)$. This scalar operation will avoid the N multiplications of (11.4.22) and reduce the overall computations by approximately $O(N)$.

BPE residual update

One more update needed to compute the gain filter in (11.4.9) is the BPE residual. To update $\varepsilon^b(n)$, make the substitutions $\mathbf{U} = \mathbf{X}_{0,N-1}(n)$, $\mathbf{v} = \pi(n)$, and $\mathbf{z} = \mathbf{y} = z^{-N}\mathbf{x}(n)$ in (9.4.5b). An approach similar to the one leading to (11.4.15) for the FPE residual produces:

$$\varepsilon^b(n) = \varepsilon^b(n-1) + \frac{[e^b(n|n)]^2}{\gamma_N(n)}, \tag{11.4.23}$$

and substitution of (11.4.21) in the above gives the very efficient update

$$\varepsilon^b(n) = \varepsilon^b(n-1) + e^b(n|n)e^b(n|n-1). \tag{11.4.24}$$

Angle update

The angle parameter update $\gamma_N(n)$ is also required to complete the set of backward error updates. As was done for the gain filter, this is achieved by computing $\gamma_{N+1}(n)$, the optimal angle update for the $N+1$ length filter, using two different approaches, and then equating the results. This procedure will then produce a derivation of $\gamma_N(n)$ in terms of $\gamma_N(n-1)$. Therefore, from the definition (11.2.34),

$$\gamma_{N+1}(n) = 1 - \langle \mathbf{x}_{N+1}(n), \mathbf{g}_{N+1}(n) \rangle. \tag{11.4.25}$$

Using (11.4.7) for $\mathbf{g}_{N+1}(n)$ and partitioning $\mathbf{x}_{N+1}(n)$ into

$$\mathbf{x}_{N+1}^T(n) = [x(n), \mathbf{x}_N^T(n-1)],$$

it is therefore easy to derive

$$\gamma_{N+1}(n) = \gamma_N(n-1) - \frac{[e^f(n|n)]^2}{\varepsilon^f(n)}. \tag{11.4.26}$$

Then, using (11.4.17) for $e^f(n|n)$ in (11.4.26) gives the efficient update

$$\gamma_{N+1}(n) = \gamma_N(n-1)\frac{\varepsilon^f(n-1)}{\varepsilon^f(n)}. \tag{11.4.27}$$

Using a different partitioning of $\mathbf{x}_{N+1}(n)$ then produces the other desired relation for $\gamma_{N+1}(n)$. Let $\mathbf{x}_{N+1}(n)$ now be partitioned into

$$\mathbf{x}_{N+1}^T(n) = [\mathbf{x}_N^T(n), x(n-N)],$$

and use (11.4.8) for $\mathbf{g}_{N+1}(n)$. Simplifying the result then produces

$$\gamma_{N+1}(n) = \gamma_N(n) - \frac{[e^b(n|n)]^2}{\varepsilon^b(n)}, \tag{11.4.28}$$

and using (11.4.23) then produces the resulting update

$$\gamma_N(n) = \gamma_{N+1}(n)\frac{\varepsilon^b(n)}{\varepsilon^b(n-1)}. \tag{11.4.29}$$

However, there is another form that allows $\gamma_N(n)$ to be computed without a knowledge of the present backward residual. To find this, divide (11.4.24) by $\varepsilon^b(n)$ and invert, giving

$$\frac{\varepsilon^b(n)}{\varepsilon^b(n-1)} = [1 - k(n)e^b(n|n-1)]^{-1}, \tag{11.4.30}$$

where (11.4.9) has also been used for $k(n)$. Using (11.4.30) in (11.4.29) thus gives

$$\gamma_N(n) = [1 - k(n)e^b(n|n-1)]^{-1}\gamma_{N+1}(n). \tag{11.4.31}$$

The benefit of this form is that it allows $\gamma_N(n)$ to be computed earlier in the FTF recursion than would the form (11.4.29). This, in turn, leads to a reduction in the required computations if (11.4.31) is used.

At this point, all the derivations have been completed in order to implement the basic FTF algorithm. For easy reference, the required equations are listed in appropriate order in Table 11.1. The completed derivation of the gain update in (11.4.42) is left as an exercise. A computational count of the method in Table 11.1 shows that approximately $8N$ arithmetic operations (i.e., multiplications and adds) are required to implement the general LS joint process prediction (i.e., to update $\mathbf{w}_N(n)$). If only the linear prediction filter needs to be updated, then approximately $6N$ operations are necessary. This algorithm is referred to as the "basic" FTF algorithm, since it is suggested rather naturally by the projection method derivation.

One of the prime benefits of the FTF is that its convergence speed is insensitive to the correlation properties of the data. This is in contrast to the gradient-based adaptive methods such as LMS, which have adaptation speeds that are very dependent on the correlation properties of the data. As an example, consider the problem of systems identification in which it is not feasible to use an uncorrelated signal as input to the system. The telephone echo cancellation problem of Chapter 6 becomes this type of problem when the echo canceller must adapt using the highly correlated input speech. Suppose the second-order autoregressive (AR) signal of Figure 11.2 is used as the input $x(n)$ to an unknown and $d(n)$ is the system output. This AR signal

Table 11.1 Basic FTF Algorithm

Initialization:

$$\mathbf{b}_N(0) = \mathbf{f}_N(0) = \mathbf{w}_N(0) = \mathbf{g}_N(0) = 0,$$

$$\gamma_N(0) = 1.0; \qquad \varepsilon^f(0) = \varepsilon^b(0) = \delta, \text{ small positive constant.}$$

For $n = 1$ to n _final_ do:

$$e^f(n|n - 1) = x(n) - \mathbf{x}_N^T(n - 1)\mathbf{f}_N(n - 1), \tag{11.4.32}$$

$$e^f(n|n) = \gamma_N(n - 1)e^f(n|n - 1), \tag{11.4.33}$$

$$\varepsilon^f(n) = \varepsilon^f(n - 1) + e^f(n|n)e^f(n|n - 1), \tag{11.4.34}$$

$$\mathbf{f}_N(n) = \mathbf{f}_N(n - 1) + e^f(n|n - 1)\mathbf{g}_N(n - 1), \tag{11.4.35}$$

$$\gamma_{N+1}(n) = \frac{\varepsilon^f(n - 1)}{\varepsilon^f(n)}\gamma_N(n - 1), \tag{11.4.36}$$

$$\begin{bmatrix} \mathbf{k}_N(n) \\ k(n) \end{bmatrix} = \begin{bmatrix} 0 \\ \mathbf{g}_N(n - 1) \end{bmatrix} + \frac{e^f(n|n)}{\varepsilon^f(n)}\begin{bmatrix} 1 \\ -\mathbf{f}_N(n) \end{bmatrix}, \tag{11.4.37}$$

$$e^b(n|n - 1) = x(n - N) - \mathbf{x}_N^T(n)\mathbf{b}_N(n - 1), \tag{11.4.38}$$

$$\gamma_N(n) = [1 - k(n)e^b(n|n - 1)]^{-1}\gamma_{N+1}(n), \tag{11.4.39}$$

$$e^b(n|n) = \gamma_N(n)e^b(n|n - 1), \tag{11.4.40}$$

$$\varepsilon^b(n) = \varepsilon^b(n - 1) + e^b(n|n)e^b(n|n - 1), \tag{11.4.41}$$

$$\mathbf{g}_N(n) = [\mathbf{k}_N(n) + k(n)\mathbf{b}_N(n - 1)]\frac{\gamma_N(n)}{\gamma_{N+1}(n)}, \tag{11.4.42}$$

$$\mathbf{b}_N(n) = \mathbf{b}_N(n - 1) + \mathbf{g}_N(n)e^b(n|n - 1). \tag{11.4.43}$$

Joint process extension:

$$e(n|n - 1) = d(n) - \mathbf{x}_N^T(n)\mathbf{w}_N(n - 1), \tag{11.4.44}$$

$$\mathbf{w}_N(n) = \mathbf{w}_N(n - 1) + \mathbf{g}_N(n)e(n|n - 1). \tag{11.4.45}$$

---

was generated by passing an uncorrelated unity power signal through the IIR filter with poles at $(z_1, z_2) = 0.9e^{\pm j\pi/6}$. Since this signal is highly correlated, the convergence of the LMS algorithm should be affected considerably. For this example, the system was chosen to have the two non-zero coefficients $(h_0, h_1) = (0.2, 0.7)$. Therefore, the system output $d(n)$ is given by

$$d(n) = 0.2x(n) + 0.7x(n - 1),$$

and the coefficients of the adaptive transversal filter should converge to 0.2 and 0.7.

In Figure 11.3, examples of using the LMS filter to identify the system are

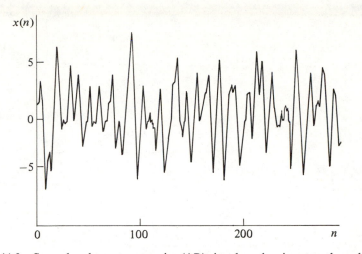

Figure 11.2   Second-order autoregressive (AR) signal used as input to the unknown system.

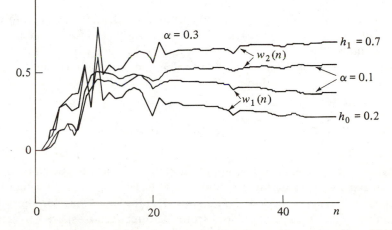

Figure 11.3   Adaptation properties of the LMS algorithm for systems identification using correlated system input signal.

shown. As can be seen, the LMS algorithm is very sluggish in converging to the correct parameters. This behavior is exactly as predicted by the analytical results of Chapter 5. The value of $\alpha = 0.3$ was the largest attained before divergence. On the other hand, the results of using the FTF method to identify the system having the AR input signal is shown in Figure 11.4. The FTF converges extremely rapidly, independent of the correlation properties of the input signal. The initialization parameter δ in Table 11.1 was set equal to unity. In general, the FTF performance is not very sensitive to the value of δ

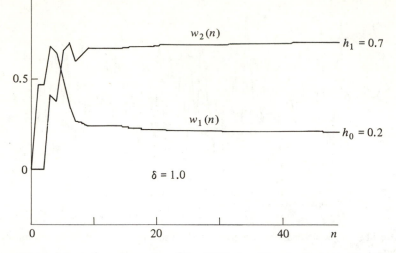

Figure 11.4   Adaptation properties of the FTF algorithm for systems identification using correlated system input signal.

selected. The LMS algorithm could only produce results similar to Figure 11.4 if the system input were uncorrelated and the LMS gain were very high.

The number of arithmetic operations involved in updating the FTF may be decreased by approximately $O(N)$ by redefining the gain filter slightly. This is explored in the next section.

11.5 Further Computational Reductions

There is a further reduction of approximately $O(N)$ arithmetic operations, which occurs by considering a variant of the basic FTF algorithm that will be called the *gain-normalized FTF*. However, the equations for the gain-normalized FTF are not as obvious from the direct vector space derivation as is the basic FTF of Table 11.1 For this reason, it is developed separately in this section.

The basic approach is to normalize the gain vector by the associated angle parameter; that is, define new gain vectors as

$$\mathbf{c}_N(n) = \frac{\mathbf{g}_N(n)}{\gamma_N(n)}, \tag{11.5.1a}$$

$$\mathbf{c}_N(n-1) = \frac{\mathbf{g}_N(n-1)}{\gamma_N(n-1)}, \tag{11.5.1b}$$

$$\mathbf{c}_{N+1}(n) = \frac{\mathbf{g}_{N+1}(n)}{\gamma_{N+1}(n)}. \tag{11.5.1c}$$

Using (11.4.27), it is easy to derive that

$$\frac{\mathbf{g}_N(n-1)}{\gamma_{N+1}(n)} = \mathbf{c}_N(n-1)\frac{\varepsilon^f(n)}{\varepsilon^f(n-1)}, \tag{11.5.2}$$

and using (11.4.16) and (11.4.29), it is straightforward to show

$$\frac{e^f(n|n)}{\varepsilon^f(n)\gamma_{N+1}(n)} = \frac{e^f(n|n-1)}{\varepsilon^f(n-1)}. \tag{11.5.3}$$

Dividing (11.4.7) by $\gamma_{N+1}(n)$ and using (11.5.3), (11.5.1c) then gives

$$\mathbf{c}_{N+1}(n) = \begin{bmatrix} 0 \\ \dfrac{\varepsilon^f(n)}{\varepsilon^f(n-1)}\mathbf{c}_N(n-1) \end{bmatrix} + \frac{e^f(n|n-1)}{\varepsilon^f(n-1)}\begin{bmatrix} 1 \\ -\mathbf{f}_N(n) \end{bmatrix}. \tag{11.54}$$

Analogous to the preceding, both sides of (11.4.8) may also be divided by $\gamma_{N+1}(n)$. Substituting (11.4.29) for $\gamma_{N+1}(n)$ into the result and simplifying somewhat then provides

$$\mathbf{c}_{N+1}(n) = \begin{bmatrix} \mathbf{m}_N(n) \\ m(n) \end{bmatrix} = \begin{bmatrix} \dfrac{\varepsilon^b(n)}{\varepsilon^b(n-1)}\mathbf{c}_N(n) \\ 0 \end{bmatrix} + \frac{e^b(n|n)}{\varepsilon^b(n-1)\gamma_N(n)}\begin{bmatrix} -\mathbf{b}_N(n) \\ 1 \end{bmatrix}, \tag{11.5.5}$$

where $\mathbf{m}_N(n)$ is the vector of the first N components of $\mathbf{c}_{N+1}(n)$. The bottom partition of (11.5.5) defines the scalar $m(n)$ as

$$m(n) = \frac{e^b(n|n)}{\varepsilon^b(n-1)\gamma_N(n)}, \tag{11.5.6}$$

which will be useful in work to follow. However, $m(n)$ need not be computed using (11.5.6), since it is already available as the $N + 1$st component of $\mathbf{c}_{N+1}(n)$ from the update in (11.5.4). Therefore, the top partition in (11.5.5) gives

$$\mathbf{c}_N(n) = \frac{\varepsilon^b(n-1)}{\varepsilon^b(n)}[\mathbf{m}_N(n) + m(n)\mathbf{b}_N(n)]. \tag{11.5.7}$$

As in Section 11.4, this recursion is in terms of $\mathbf{b}_N(n)$, which has not yet been computed. Therefore, substitute (11.4.43) into (11.5.7), providing

$$\mathbf{c}_N(n) = \frac{\varepsilon^b(n-1)}{\varepsilon^b(n) - \varepsilon^b(n-1)m(n)e^b(n|n)}[\mathbf{m}_N(n) + m(n)\mathbf{b}_N(n-1)]. \tag{11.5.8}$$

However, using (11.4.23) and (11.5.6), it is straightforward to show that

$$\frac{\varepsilon^b(n-1)}{\varepsilon^b(n) - \varepsilon^b(n-1)m(n)e^b(n|n)} = 1, \tag{11.5.9}$$

and, therefore, (11.5.8) becomes

$$\mathbf{c}_N(n) = \mathbf{m}_N(n) + m(n)\mathbf{b}_N(n-1). \tag{11.5.10}$$

An update for $\gamma_N(n)$ results from first solving (11.5.9) for the ratio of BPE residuals:

$$\frac{\varepsilon^b(n)}{\varepsilon^b(n-1)} = 1 + m(n)e^b(n|n).$$ (11.5.11)

Substituting (11.5.11) and (11.4.21) into (11.4.29) then gives

$$\gamma_N(n) = [1 - \gamma_{N+1}(n)m(n)e^b(n|n-1)]^{-1}\gamma_{N+1}(n).$$ (11.5.12)

The main computational savings of the gain-normalized FTF is that the BPE $e^b(n|n-1)$ need not be computed using the $O(N)$ operations of (11.4.38), but instead may be computed by a simple scalar multiply and divide. To see this, substitute (11.5.11) into (11.4.29) and solve the resulting equation for $e^b(n|n-1)$. After simplification, this gives

$$e^b(n|n-1) = \frac{1}{m(n)}\left[\frac{1}{\gamma_{N+1}(n)} - \frac{1}{\gamma_N(n)}\right].$$ (11.5.13)

Then, substituting (11.5.6) in (11.5.13) and using (11.4.29) produces

$$e^b(n|n-1) = \left[\frac{\varepsilon^b(n) - \varepsilon^b(n-1)}{e^b(n|n)\varepsilon^b(n-1)}\right]\varepsilon^b(n-1).$$ (11.5.14)

However, using (11.5.9), it is easy to show that the term in brackets in (11.5.14) equals $m(n)$, and therefore

$$e^b(n|n-1) = m(n)\varepsilon^b(n-1).$$ (11.5.15)

The remaining equations in the gain-normalized FTF are very similar to those in the basic FTF, and the entire algorithm is listed in Table 11.2 for easy reference. The main difference is that the transversal filter updates are done with the "updated" error and "normalized" gain vector, and the comuptation of $e^b(n|n-1)$ no longer requires $O(N)$ computations.

PROBLEMS

1. Show that the $(N+1) \times n$ transversal filter operator $\mathbf{K}_{0,N-1,\pi}(n)$ may be partitoned as

$$\mathbf{K}_{0,N-1,\pi}(n) = \begin{bmatrix} \mathbf{K}_{0,N-1}(n-1) & \mathbf{0}_N \\ \mathbf{c}_{n-1}^T & 1 \end{bmatrix}.$$

2. Show that the vector $\mathbf{c}_{n-1}$ in P1 above lies in the subspace $\{\mathbf{X}_{0,N-1}(n-1)\}$.

3. Suppose that the N coefficient transversal filter operator $\mathbf{K}_\mathbf{U}$ has been computed based upon a subspace $\{\mathbf{U}\}$, where $\mathbf{U}$ is an $n \times N$ data matrix. Show that if an $n \times 1$ vector $\mathbf{v}$ is appended as the *first* column of the $n \times (N+1)$ matrix $[\mathbf{v}, \mathbf{U}]$, then the resulting update for $\mathbf{K}_{\mathbf{vU}}$ is given by (11.3.9b).

4. Given that

$$\mathbf{g}_N(n-1) = \mathbf{K}_{1,N}(n)\pi(n),$$

Table 11.2 Gain-Normalized FTF Algorithm

Initialization:

$$\mathbf{b}_N(0) = \mathbf{f}_N(0) = \mathbf{w}_N(0) = \mathbf{c}_N(0) = 0,$$

$$\gamma_N(0) = 1.0; \qquad \varepsilon^f(0) = \varepsilon^b(0) = \delta, \text{ small positive constant.}$$

For $n = 1$ to n *final* do:

$$e^f(n|n-1) = x(n) - \mathbf{x}_N^T(n-1)\mathbf{f}_N(n-1), \tag{11.5.16}$$

$$e^f(n|n) = \gamma_N(n-1)e^f(n|n-1), \tag{11.5.17}$$

$$\varepsilon^f(n) = \varepsilon^f(n-1) + e^f(n|n)e^f(n|n-1), \tag{11.5.18}$$

$$\mathbf{f}_N(n) = \mathbf{f}_N(n-1) + e^f(n|n)\mathbf{c}_N(n-1), \tag{11.5.19}$$

$$\gamma_{N+1}(n) = \frac{\varepsilon^f(n-1)}{\varepsilon^f(n)}\gamma_N(n-1), \tag{11.5.20}$$

$$\begin{bmatrix} \mathbf{m}_N(n) \\ m(n) \end{bmatrix} = \begin{bmatrix} 0 \\ \mathbf{c}_N(n-1) \end{bmatrix} + \frac{e^f(n|n)}{\varepsilon^f(n-1)}\begin{bmatrix} 1 \\ -\mathbf{f}_N(n-1) \end{bmatrix}, \tag{11.5.21}$$

$$e^b(n|n-1) = m(n)\varepsilon^b(n-1), \tag{11.5.22}$$

$$\gamma_N(n) = [1 - \gamma_{N+1}(n)m(n)e^b(n|n-1)]^{-1}\gamma_{N+1}(n), \tag{11.5.23}$$

$$e^b(n|n) = \gamma_N(n)e^b(n|n-1), \tag{11.5.24}$$

$$\varepsilon^b(n) = \varepsilon^b(n-1) + e^b(n|n)e^b(n|n-1), \tag{11.5.25}$$

$$\mathbf{c}_N(n) = \mathbf{m}_N(n) + m(n)\mathbf{b}_N(n-1), \tag{11.5.26}$$

$$\mathbf{b}_N(n) = \mathbf{b}_N(n-1) + \mathbf{c}_N(n)e^b(n|n). \tag{11.5.27}$$

Joint process extension:

$$e(n|n-1) = d(n) - \mathbf{x}_N^T(n)\mathbf{w}_N(n-1), \tag{11.5.28}$$

$$e(n|n) = \gamma_N(n)e(n|n-1), \tag{11.5.29}$$

$$\mathbf{w}_N(n) = \mathbf{w}_N(n-1) + \mathbf{c}_N(n)e(n|n). \tag{11.5.30}$$

show that

$$\mathbf{g}_N(n) = \mathbf{K}_{0,N-1}(n)\pi(n).$$

5. Given that

$$\gamma_m(n-1) = \langle \pi(n), \mathbf{P}_{1,m}^{\perp}(n)\pi(n) \rangle,$$

show that

$$\gamma(n) = \langle \pi(n), \mathbf{P}_{0,m-1}^{\perp}(n)\pi(n) \rangle.$$

6. Verify the update equation (11.4.19) for the BPE filter by using $\mathbf{U} = \mathbf{X}_{0,N-1}(n)$ and $\mathbf{v} = z^{-N}\mathbf{x}(n)$ in (11.3.9a).

7. Verify that the substitutions $\mathbf{U} = \mathbf{X}_{0,N-1}(n)$, $\mathbf{v} = \pi(n)$, and $\mathbf{z} = \mathbf{y} = z^{-N}$ lead to the BPE residual (11.4.15).

8. Given a time signal $v(n) = \{v(1), v(2), v(3), \dots\} = \{4, 2, 4, \dots\}$, compute the following:
   (a) The vectors $\mathbf{v}(2)$ and $\mathbf{v}(3)$.
   (b) The vectors $z^{-1}\mathbf{v}(2)$ and $z^{-2}\mathbf{v}(2)$.
   (c) The vectors $z^{-1}\mathbf{v}(3)$ and $z^{-2}\mathbf{v}(3)$.
   Let $\mathbf{u}(n) = z^{-1}\mathbf{v}(n)$ in the rest of this problem. Do the following:
   (d) Compute $\mathbf{P}_\mathbf{u}(2)$.
   (e) Compute $\mathbf{P}_\mathbf{u}(3)$.
   (f) Compute the LS predictions of $\mathbf{v}(n)$ using $\mathbf{u}(n)$ for $n = 2, 3$.
   (g) Compute the error vectors $\mathbf{e}_1^f(2)$, $\mathbf{e}_1^f(3)$.
   (h) Draw the vector spaces $\{\mathbf{u}(2)\}$ and $\{\mathbf{u}(3)\}$.

9. Let $\mathbf{u}(3) = [3, 2, 3]^T$ and let $\pi(3) = [0, 0, 1]$.
   (a) Show that $\mathbf{P}_{\mathbf{u},\pi}(3)$ can be written in the form

$$\mathbf{P}_{\mathbf{u},\pi}(3) = \begin{bmatrix} \mathbf{P}_\mathbf{u}(2) & \mathbf{0}_2 \\ \mathbf{0}_2^T & 1 \end{bmatrix}.$$

   (b) Compute the vector $\mathbf{g} = \mathbf{P}_\mathbf{u}(3)\pi(3)$.
   (c) Finish this sentence (in words): "$\mathbf{g}$ is ... ."
   (d) Draw a picture that denotes $\pi(3)$, $\{\mathbf{u}(3), \pi(3)\}$, $\mathbf{u}(3)$, $\{\mathbf{u}(3)\}$, and $\mathbf{g}$.

REFERENCES

1. L. Ljung, M. Morf, and D.D. Falconer, "Fast Calculation of Gain Matrices for Recursive Estimation Schemes," *Int. J. of Control*, vol. 27, pp. 1–17, January 1978.
2. D.D. Falconer and L. Ljung, "Application of Fast Kalman Estimation to Adaptive Equalization," *IEEE Trans. on Communications*, vol. COM-26, pp. 1439–1445, October 1978.
3. G. Carayannis, D. Manolakis, and N. Kalouptsidis, "A Fast Sequential Algorithm for Least Squares Filtering and Prediction," *IEEE Trans. on Acous., Speech, and Signal Processing*, vol. ASSP-31, pp. 1394–1403, December 1983.
4. J.M. Cioffi, "Fast Transversal Filter Applications for Communications Applications." Ph. D. Dissertation, Stanford University, 1984.
5. J.M. Cioffi and T. Kailath, "Fast Recursive-Least-Squares Transversal Filters for Adaptive Filtering," *IEEE Trans. on Acous., Speech, and Signal Processing*, vol. ASSP-32, pp. 304–338, April 1984.
6. M.L. Honig, "Recursive, Fixed-Order Covariance Least Squares Algorithms," *Bell Sys. Tech. J.*, vol. 62, pp. 2961–2992, December 1983.
7. J.M. Cioffi and T. Kailath, "An Efficient RLS Data-Driven Echo Canceller for Fast Initialization of Full Duplex Data Transmission," *IEEE Trans. on Communications*, vol. COM-33, pp. 601–611, July 1985.
8. J.M. Cioffi and T. Kailath, "Windowed Fast Transversal Filters Adaptive Algorithms with Normalization," *IEEE Trans. on Acous., Speech, and Signal Processing*, vol. ASSP-33, pp. 607–625, June 1985.
9. J.M. Cioffi, "When Do I Use an RLS Adaptive Filter?" *Proceedings*, 19th Asilomar Conf. on Circuits, Systems, and Computers, Pacific Grove, CA, November 1985.

Index